PLANT CYTOGENETICS, BREEDING AND EVOLUTION

PLANT CYTOGENETICS, BREEDING AND EVOLUTION

M.S. Rana

CYBER TECH PUBLICATIONS
4264/3, Ansari Road, Daryaganj, New Delhi-110002 (India)
Ph.: 011-23244078, 011-43559448 Fax: 011-23280028
E-Mail: cyberpublicationsdelhi@yahoo.com
Website: www.cybertechpublications.com

PLANT CYTOGENETICS, BREEDING AND EVOLUTION

M.S. Rana

First Edition 2010

Published by :

G.S. Rawat for **Cyber Tech Publications**
4264/3, Ansari Road, Daryaganj, New Delhi-110002 (India)
Ph.: 011-23244078, 011-43559448 Fax: 011-23280028
E-Mail: cyberpublicationsdelhi@yahoo.com
Website: www.cybertechpublications.com

Printed at: Asian Offset Printers, Delhi-53

Preface

Plant cytogenetics comprises a topic of broad interest and increasing importance in plant science. In keeping with the exciting advances in plant genetics and genomics, we believe that a comprehensive and up-to-date reference on plant cytogenetics would be of great interest and value for researchers, instructors, and students with interests in genetics, plant biology, and plant genomics. The field of cytogenetics is currently expanding at the same fast pace that genome technologies are moving but no reliable reference exists that captures the current state of progress. The goal then of the proposed reference is to provide a scholarly and timely review of plant cytogenetics that covers traditional aspects of plant cytogenetics together with the emerging and exciting new areas of genomics, epigenetics, evolution, and chromosome engineering. Advances in plant genetics and genomics are driving major changes in how we view fundamental concepts such as gene regulation, genome organization, and plant evolution.

Plant Cytogenetics: Breeding and Evolution Chromosome Function provides a unique and critical perspective on these precepts. The genome revolution has opened new avenues for understanding and modifying plant genomes. A modern reference book on plant cytogenetics will be a major asset as plant geneticists move from engineering genes and pathways to engineering more complex genetic elements such as whole chromosomes and composite genomes.

We expect that this definitive reference would serve just such a role covering the fundamentals of plant cytogenetics on this topic of great historic and contemporary importance.

—*Author*

CONTENTS

1

INTRODUCTION

BASIC TERMINOLOGY

Chromosome

A molecule of DNA, and associated proteins bound together. Each chromosome contains:

Centromere: specialized region that is the last part of a replicated chromosome to disjoin at meiosis or mitosis

Kinetochore: protein structure surrounding centromere that coordinates spindle attachment and chromosome movement

Telomere: specialized regions at both ends of the chromosome

Euchromatin: stains normally, gene rich region

Heterochromatin: stains darkly, gene poor region

What composes telomeres and centromeres at the molecular level?

Telomeres are composed of short tandem repeats of DNA that are added to chromosome ends by the enzyme telomerase. These repeat sequences are fairly well conserved across all eukaryotes, and the repeat sequence TTTAGGG is conserved in most plant species (Gill and Friebe, 1998).

Centromeres also contain blocks of repeated DNA sequences, but these sequences seem much less conserved across species. Centromere regions range from about 1 Mbp (Arabidopsis) to more than 100 Mbp (wheat) in size (Bennetzen 1998). Richards and Dawe (1998) suggested that "centromere DNA sequences are not always sufficient, nor are they always necessary for specification of centromere activity. The observation that there is little sequence conservation among the known centromere repeats…is consistent with the idea that primary DNA sequence may provide only part of the information necessary to specify centromere function. The remaining information is thought to be epigenetic, superimposed on the primary nucleotide information, in the form of stable, self-propagating chromatin states."

This suggests that DNA sequence analysis alone will not tell us too much about what a centromere is! The centromere seems to be the result of an interaction between various DNA sequences and the chromosome-associated proteins. Interestingly, although centromeric DNA sequences seem to be highly variable, one of the key proteins they interact with to produce kinetochores is a highly conserved protein called CenH3, which is a variant histone protein. This protein is found in yeast, insects, vertebrates, and plants. There may be other centromeric proteins that are rapidly evolving, however, to maintain the interaction between the conserved $CenH_3$ and the rapidly changing centromeric DNA sequences.

The most detailed analyses of plant centromere DNA sequences have been in rice. Cultivated rice (*Oryza sativa*) centromeres contain two domains: a 155 bp sequence called CentO that is tandemly repeated many times (and called "satellite" DNA because most such repeats have different bulk density than the rest of the genome and can be separated on density gradients), and a set of retrotransposon-derived sequences, called CRR, that are dispersed among these repeats and also found in the centromeres of other grass species.

Lee et al. (2005) analyzed the sequences of centromeres from 17 different species in the *Oryza* genus and found that they did not all share a common centromere repeat sequence. Some of the wild species have quite different centromeric repeats, indicating the rapid evolution of the sequences that can occur. Even more surprising is the result from Nasuda et al. (2005) that artificially broken barley chromosomes that completely lack the centromeric associated repeats can still function by organizing kinetochore formation. How are centromeres defined genetically?

Chromatid: One molecule of the double helix of DNA in a replicated chromosome.

Following DNA replication and before meiosis or mitosis, each chromosome has two chromatids joined at the centromere.

Homologous chromosomes: Chromosomes that carry the same genes in the same order and that pair at meiosis.

May have different alleles at common genes.

Homoeologous (homeologous) chromosomes: Chromosomes that carry some, but not all, of the same genes (or related genes). They do not pair at meiosis, or pair with low frequency.

Genome: All the genetic information (DNA, chromosomes) in a cell of an organism. Often the nuclear genome is implied, but don't forget about the mitochondrial and chloroplast genomes!

Locus: A particular location on a chromosome. It may refer to a gene that codes for a protein (e.g., the waxy gene in corn). Or it may refer to a region of the chromosome that does not code for a protein (e.g., UMC187, a corn RFLP locus identified by a genomic clone).

So, all genes are loci, but not all loci are genes!

Allele: Alternate form a locus.

Designated by *A*, *a* – implying dominance of the *A* allele over the *a* allele. Or by *A*1, *A*2, …, *A*n to account for multiple alleles without implying dominance relationships.

Breeding works by selecting for different alleles, not different genes!

Nomenclature of chromosome numbers and DNA content.

n = gametophytic chromosome complement

2n = sporophytic chromosome complement

x = number of chromosomes in one complete homoeologous set of chromosomes.

2x = number of chromosomes in a diploid sporophyte

4x = number of chromosomes in a tetraploid sporophyte. etc.

Examples

Diploid Maize 2n = 2x = 20

Triploid Banana 2n = 3x = 33

Tetraploid Alfalfa 2n = 4x = 32

Hexaploid Wheat 2n = 6x = 42

Octoploid Sugarcane 2n = 8x = 80

So, most cells in a maize plant have 2n = 2x = 20 chromosomes, but maize pollen cells each have n = x = 10 chromosomes.

Compare: Relative amount of DNA in a cell

C = "constant" – the relative amount of DNA in a gametophytic cell

2C = relative amount of DNA in a sporophytic cell before DNA replication

4C = relative amount of DNA in a sporophytic cell after DNA replication and before mitosis/cell division.

Absolute amount of DNA in a cell – "C-value"

Given in picograms (10-12 g)

1 pg = 0.965 x 1012 nucleotides (base pairs)

See Arumunagathan and Earle (1991) for C-values for many plant species.

Euploidy

Haploid: A sporophytic plant having the gametophytic chromosome number (*n*), and derived directly from a gametophyte (without fertilization).

If the species is diploid ($2n = 2x$), a haploid of that species has a single set of the chromosome ($n = x$).

If the species is polyploid ($2n > 2x$), a haploid of that species has more than a single set of the chromosomes ($n > x$), but is still called a haploid. For example, a haploid derived from tetraploid alfalfa ($2n = 4x = 32$) has two copies of the basic chromosome set ($n = 2x = 16$)

Doubled Haploid: Haploidy can be induced in some species (e.g., in barley, tobacco with some regularity; in maize, alfalfa, wheat, oat, potato it is done with greater difficulty). This can be accomplished by anther/pollen culture, interspecific crosses, or with some reproductive mutants. Doubling is achieved with colchicine or by tissue culture techniques.

Why make doubled haploids? Since haploids have only one allele per locus (if from a diploid species), doubled haploids are immediately completely homozygous at every locus. This may be useful for breeding purposes (homozygosity in one generation vs. selfing for many generations) and for genetics applications (linkage mapping is simplified).

Polyploid: A species having more than two basic sets of chromosomes per sporophytic cell. Many species are polyploids (70% of grasses, 23% of legumes, for example).

Types of polyploids (Stebbins): Nomenclature: A represents the set of chromosomes in the A "genome." B represents the set of chromosomes in the B genome, etc.

Autopolyploids: many sets of a single genome: AA 6 AAAA

Allopolyploids: two or more distinct genomes: AA + BB 6 AABB or AB 6 AABB. (Note that A and B can be homoeologous sets of chromosomes; i.e., they are similar sets of chromosomes, but A chromosomes do not pair with B chromosomes at meiosis.)

Segmental allopolyploids: genomes are somewhat differentiated, but pairing of chromosomes from different genomes is still possible: A1A1 + A2A2 =A1A1A2A2.

Autoallopolyploids: a mixture of common and distinct genomes.

Pairing of chromosomes at meiosis:

Bivalents (II) – two chromosomes pair at meiosis I

Trivalents (III) – three chromosomes pair at meiosis I

Quadrivalents (IV) – four chromosomes pair at meiosis I

Allopolyploids do not have pairing higher than bivalents.

Autopolyploids may have pairing higher than bivalents. In some cases, autopolyploids have only bivalents, but the crucial distinction is that any of the more than two homologous chromosomes can pair, there is no restriction of which two homologs pair.

Because the terms auto- and alloploidy imply an evolutionary history that is often not really known, the terms disomic and polysomic polyploids are preferred. These terms are based solely on the pairing relationships among the chromosomes, and do not imply the evolutionary origin of the polyploidy species.

Disomic polyploid: only bivalents occur at meiosis, and each chromosome can pair with only one other chromosome in the genome.

Polysomic polyploid: each chromosome can pair with more than one other chromosome in the genome; groups of more than two chromosomes may pair at meiosis.

Amphiploid: a man-made polyploid synthesized by doubling the chromosomes of an interspecific hybrid. Some examples include Triticale (wheat x rye) and Raphanobrassica (radish x cabbage).

Some examples to clarify:

Maize: 2n = 2x = 20. The basic chromosome set has 10 distinct chromosomes.

Each sporophytic cell has 2 copies of each chromosome.

Wheat: 2n = 6x = 42. The basic chromosome set of wheat has 7 distinct chromosomes. Bread wheat has three related genomes, A, B, and D.

Chromosome 1 of genome A is structurally similar to chromosome 1 of genome B and chromosome 1 of genome D, chromosome 1 of genome A only pairs with chromosome 1 of genome A!

Alfalfa: 2n = 4x = 32. The basic chromosome set of alfalfa has 8 chromosomes.

Cultivates alfalfa cells each have four copies of each of the 8 chromosomes. Any one of these four copies of a common homolog can pair with any of the other three at meiosis, even though quadrivalents rarely form.

Now that you grasp the basic rules, it is time to learn that in reality, the distinction between diploidy and polyploidy can be fuzzy. Many diploid species are in fact "paleopolyploids" meaning that they have many duplicated genes, implying a polyploidy ancestor in their evolutionary history, but through evolution, the species developed regular disomic inheritance.

Paleopolyploidy was discovered originally with molecular linkage maps (in corn and soybean). A single RFLP probe often identifies more than one locus in these species, and conserved linkage arrangements among some of the duplicated loci ("large-scale duplications") were observed. Whereas a single duplicated locus is not evidence for ancient polyploidy, larger stretches of multiple duplicated linkage arrangements are! More recently, DNA sequencing has provided additional evidence that paleopolyploidy is common. The complete DNA sequence of *Arabidopsis*, which has one of the simplest regular diploid genomes among plants, reveals large scale duplication events within the genome, suggesting ancient polyploidy.

The result of this large-scale duplication is that *genetic redundancy* is common in plants:

More than one gene may code for similar functions.

"Knocking out" even a simple phenotype may be difficult if more than one locus affects the phenotype.

Duplicate epistasis may be common in plants.

But, if the duplications are sufficiently ancient, the duplicated genes may evolve new functions or function under different conditions or developmental stages.

Meiosis

Compare diploid, allotetraploid, autotetraploid for one set of homologous or homoeologous chromosomes

Polyploid segregation ratios.

What are the expected segregation ratios of a single locus in a tetrasomic polyploid?

They depend on the recombination frequency between the locus and the centromere. As an example, we demonstrate segregation ratios at three hypothetical loci on, *A*, *B*, and *C*, that differ in their recombination frequencies from the centromere. The *A* locus is tightly linked to the centromere, so that no recombination occurs between the locus and the centromere. The A locus segregates exactly like the centromeres themselves. At meiosis I, two homologous centromeres move to one pole, and the other two centromeres move to the other pole. So, if the plant is tetra-allelic, A1A2A3A4, then at meiosis I (following DNA replication and chromosome doubling), the following segregations are possible (and equally likely):

MI

A1A1/A2A2 + A3A3/A4A4

A1A1/A3A3 + A2A2/A4A4

A1A1/A4A4 + A2A2/A3A3

At meiosis II, one copy of each centromere and one allele of each locus in parent cell move into each daughter cell:

MI MII

A1A1/A2A2 + A3A3/A4A4 A1A2 / A1A2 / A3A4 / A3A4

A1A1/A3A3 + A2A2/A4A4 A1A3 / A1A3 / A2A4 / A2A4

A1A1/A4A4 + A2A2/A3A3 A1A4 / A1A4 / A2A3 / A2A3

This results in the equal probability of the following gametic types: A1A2, A1A3, A1A4, A2A3, A2A4, A3A4 (1/6 frequency each). This mode of inheritance is called *random chromosome segregation.*

The *C* locus is far from the centromere, so that the *C* locus and the centromere are genetically unlinked. In this case, sufficient crossing-over occurs between the centromere and the *C* locus that any one allele at the *C* locus can be grouped with any set of three other alleles after the first meiotic separation – *if the chromosomes form quadrivalents.* If the *C* locus has a recombination frequency of 50% with the centromere, the gamete frequencies from genotype C1C2C3C4 are:

(1/28) C1C1

(4/28) C1C2

(4/28) C1C3

(4/28) C1C4

(1/28) C2C2

(4/28) C2C3

(4/28) C2C4

(1/28) C3C3

(4/28) C3C4

(1/28) C4C4

To figure out these gamete frequencies, realize that there is only one possible way to get a C1C1 gamete, but there are four ways to get a C1C2 gamete: the first C1 chromatid segment can be paired with either the first or the second C2 chromatid segme nt, and the second C1 chromatid segment can be paired with either the first or the second C2 chromatid segment. This reasoning applies to all of the other heterozygous gametic combinations.

When the locus exhibits these segregation ratios, it is called *random chromatid segregation.* Notice that an outcome of this type of segregation is that alleles on sister chromatids may end up in the same gamete; this result is called *double reduction.* The frequency of double reduction is measured as the frequency of gametes containing sister chromatid alleles and is denoted as a.

So, under random chromatid assortment, the frequency of double reduction = a = sum of frequencies of C1C1, C2C2, C3C3, C4C4 = 1/7. What happens if *C* has 50% recombination frequency with the centromere, but quadrivalents do not form?

Consider the gamete frequencies of a C1C2C3C4 plant with only bivalent pairing. First consider the possible pairing arrangements:

MI

(1/3) C1C1/C2C2 + C3C3/C4C4

(1/3) C1C1/C3C3 + C2C2/C4C4

(1/3) C1C1/C4C4 + C2C2/C3C3

Now, allow for recombination between centromeres and chromosomes during MI:

MI metaphase

(1/12) C1C1/C2C2 + C3C3/C4C4

(1/12) C1C2/C1C2 + C3C3/C4C4

(1/12) C1C1/C2C2 + C3C4/C3C4

(1/12) C1C2/C1C2 + C3C4/C3C4

(1/12) C1C1/C3C3 + C2C2/C4C4

(1/12) C1C3/C1C3 + C2C2/C4C4

(1/12) C1C1/C3C3 + C2C4/C2C4

(1/12) C1C3/C1C3 + C2C4/C2C4

(1/12) C1C1/C4C4 + C2C2/C3C3

(1/12) C1C4/C1C4 + C2C2/C3C3

(1/12) C1C1/C4C4 + C2C3/C2C3

(1/12) C1C4/C1C4 + C2C3/C2C3

MI metaphase MI anaphase ___

(1/12) C1C1/C2C2 + C3C3/C4C4 C1C1 / C3C3 // C2C2 / C4C4 orC1C1 / C4C4 // C2C2 / C3C3

(1/12) C1C2/C1C2 + C3C3/C4C4 C1C2 / C3C3 // C1C2 / C4C4

(1/12) C1C1/C2C2 + C3C4/C3C4 C1C1 / C3C4 // C2C2 / C3C4

(1/12) C1C2/C1C2 + C3C4/C3C4 C1C2 / C3C4 // C1C2 / C3C4

(1/12) C1C1/C3C3 + C2C2/C4C4 C1C1 / C2C2 // C3C3 / C4C4 or C1C1 / C4C4 // C2C2 / C3C3

(1/12) C1C3/C1C3 + C2C2/C4C4 C1C3 / C2C2 // C1C3 / C4C4

(1/12) C1C1/C3C3 + C2C4/C2C4 C1C1 / C2C4 // C3C3 / C2C4

(1/12) C1C3/C1C3 + C2C4/C2C4 C1C3 / C2C4 // C1C3 / C2C4

(1/12) C1C1/C4C4 + C2C2/C3C3 C1C1 / C2C2 // C3C3 / C4C4 or C1C1 / C3C3 // C2C2 / C4C4

(1/12) C1C4/C1C4 + C2C2/C3C3 C1C4 / C2C2 // C1C4 / C3C3

(1/12) C1C1/C4C4 + C2C3/C2C3 C1C1 / C2C3 // C2C3 / C4C4

(1/12) C1C4/C1C4 + C2C3/C2C3 C1C4 / C2C3 // C1C4 / C2C3

At meiosis II, one copy of each centromere and one allele of each locus in parent cell move into each daughter cell. This results in the following gametic types:

MI anaphase ___ *MII*

C1C1 / C3C3 // C2C2 / C4C4 orC1C1 / C4C4 // C2C2 / C3C3
C1C3 / C1C3 / C2C4 / C2C4 or C1C4 / C2C3 / C1C4 / C2C3

C1C2 / C3C3 // C1C2 / C4C4 C1C3 / C2C3 / C1C4 / C2C4

C1C1 / C3C4 // C2C2 / C3C4 C1C3 / C1C4 / C2C3 / C2C4

C1C2 / C3C4 // C1C2 / C3C4 C1C3 / C1C4 / C2C3 / C2C4

C1C1 / C2C2 // C3C3 / C4C4 or C1C1 / C4C4 // C2C2 / C3C3
C1C2 / C1C2 / C3C4 / C3C4 or C1C4 / C2C3 / C1C4 / C2C3

C1C3 / C2C2 // C1C3 / C4C4 C1C2 / C2C3 / C1C4 / C3C4

C1C1 / C2C4 // C3C3 / C2C4 C1C2 / C1C4 / C2C3 / C3C4

C1C3 / C2C4 // C1C3 / C2C4 C1C2 / C3C4 / C1C4 / C2C3

C1C1 / C2C2 // C3C3 / C4C4 or C1C1 / C3C3 // C2C2 / C4C4
C1C2 / C1C2 / C3C4 / C3C4 or C1C3 / C1C3 / C2C4 / C2C4
C1C4 / C2C2 // C1C4 / C3C3 C1C2 / C2C4 / C1C3 / C3C4
C1C1 / C2C3 // C2C3 / C4C4 C1C2 / C1C3 / C2C4 / C3C4
C1C4 / C2C3 // C1C4 / C2C3 C1C2 / C3C4 / C1C3 / C2C4

Summing over all the possibilities, the gametic frequencies are:

(1/6) C1C2

(1/6) C1C3

(1/6) C1C4

(1/6) C2C3

(1/6) C2C4

(1/6) C3C4

This is identical to random chromosome segregation! *So, if the chromosomes do not form quadrivalents, then loci segregate following random chromosome segregation, no matter what their position is relative to the centromere.*

To summarize: If the locus is very near the centromere, random chromosome segregation occurs.

If the locus has 50% recombination from the centromere, and quadrivalents form, random chromatid segregation occurs.

If the locus has 50% recombination from the centromere, and quadrivalents do not form, random chromosome segregation occurs.

Finally, consider a locus, B, that is not completely linked nor completely unlinked from the centromere. The frequency of double reduction at this locus will likely be somewhere between 0 and (1/7), and this will depend on the formation of quadrivalents and the probability of recombination between the centromere and locus B.

We can write general formulas for gamete frequencies of an autotetraploid as a function of a, the probability of double reduction.

Note that:

a = 0 for loci undergoing random chromosome segregation,

a = 0 if quadrivalents do not form (as in alfalfa),

a = 1/7 for loci undergoing random chromatid segregation, 11

0 < a < 1/7 for loci that are only partially linked to the centromere.

Bingham et al. (1968) estimated the frequency of double reduction in autoteraploid maize and reported that:

a = 2.2% for *wx*

a = 18% for *Sh*

They suggested that the estimate of a = 18% for *Sh* exceeded the theoretical maximum (14%) because of sampling error. In general, though, the estimates follow roughly what is expected based on the position of these loci with respect to the centromere: on chromosome 9, wx is 5 cM from the centromere, and *Sh* is about 30 cM from the centromere.

In a quadrallelic genotype, B1B2B3B4, each double reduction gamete occurs with frequency a/4. (Note that this gives 1/28 when a = 1/7 in random chromatid segregation). The other types of gametes occur at a frequency of (1 – a)/6:

Gametes from double reduction		**Gametes not from double reduction**	
Gamete	**Frequency**	**Gamete**	**Frequency**
B1B1	a/4	B1B2	(1 – a)/6
B2B2	a/4	B1B3	(1 – a)/6
B3B3	a/4	B1B4	(1 – a)/6
B4B4	a/4	B2B3	(1 – a)/6
		B2B4	(1 – a)/6
		B3B4	(1 – a)/6
Sum	a	Sum	1 – a

Genotype frequencies can then be easily derived from a Punnett square of the appropriate gamete frequencies.

Special cases: Diallelic genotypes (with two alleles per locus, B and b):

Genotype of Parent Plant

Gamete	BBBb	BBbb Frequency	Bbbb
BB	(2 + a)/4	(1 + 2a)/6	a/4
Bb	(2 – 2a)/4	(4 – 4a)/6	(2 – 2a)/4
bb	a/4	(1 + 2a)/6	(2 + a)/4

What are the genotype frequencies resulting from self-pollination of each of these three types of diallelic parents?

Offspring Genotype	Genotype of Parent Plant BBBb	BBbb	Bbbb
	Offspring genotype frequencies		
BBBB	$[(2 + \alpha)/4]^2$	$[(1 + 2\alpha)/6]^2$	$\alpha^2/16$
BBBb	$(2 + \alpha)(2 - 2\alpha)/8$	$(1 + 2\alpha)(4 - 4\alpha)/36$	$(2 - 2\alpha)\alpha/8$
BBbb	$(2+\alpha)\alpha/8+[(2-2\alpha)/4]^2$	$[(1+2\alpha)/6]^2+[(4-4\alpha)/6]^2$	$(2+\alpha)\alpha/8$ $+[(2-2\alpha)/4]^2$
Bbbb	$(2 - 2\alpha)\alpha/8$	$(1 + 2\alpha)(4 - 4\alpha)/36$	$(2 + \alpha)(2 - 2\alpha)/8$
bbbb	$\alpha^2/16$	$[(1 + 2\alpha)/6]^2$	$[(2 + \alpha)/4]^2$

Notice that if a = 0, then neither a bbbb nor a Bbbb genotype can arise from selfpollination of a BBBb genotype. Also, in this case, the genotype frequencies that arise from the self-pollination of a Bbbb plant are (1/4) BBbb, (1/2) Bbbb, (1/4) bbbb. From these results, it should be clear that homozygous genotypes are rarer in polysomic polyploids than in diploids.

Segregation Ratios in Allopolyploids

Let A represent a locus on one chromosome of a disomic tetraploid, and let B represent a locus on its homoeologous chromosome. If a plant with genotype A1A2B1B2 is selfed, the following gamete ratios result (verify this by considering the behavior of the homoeologous chromosomes at meiosis):

Gamete	Frequency
A1B1	1/4
A1B2	1/4
A2B1	1/4
A2B2	1/4

The following genotype frequencies occur in its selfed progeny:

Genotype	Frequency
A1A1B1B1	1/16
A1A1B1B2	1/8
A1A1B2B2	1/16
A1A2B1B1	1/8
A1A2B1B2	1/4
A1A2B2B2	1/8
A2A2B1B1	1/16
A2A2B1B2	1/8
A2A2B2B2	1/16

You will notice that this is simply a diploid segregation ratio with two unlinked genes. In fact, breeders treat disomic polyploids as diploids for the purposes of genetics, with the realization that many genes are likely duplicated on homoeologous chromosomes. What do the different inheritance patterns of polysomic and disomic polyploids suggest about their breeding behavior? Mac Key (1970) observed that virtually all polysomic polyploids are primarily outcrosses, whereas virtually all disomic polyploids are primarily self- fertilizing species. He proposed that the reason for this lay in their different inheritance patterns.

It seems that in most diploids and in polysomic polyploids, heterozygosity is related to vigor (this does not necessarily imply overdominant gene action, it is likely primarily due to deleterious recessive alleles at different loci in repulsion phase linkage).

Therefore, polysomic polyploids tend to be outcrossers in order to maximize heterozygosity. In contrast, disomic polyploids can "fix" heterozygosity in the homozygous state because alleles at duplicated loci on homoeologous chromosomes can interact like different alleles at the same locus in a diploid or polysomic polyploid.

Thus, using our previous example, if the A1B2 allelic interaction were favorable, it can be fixed in the A1A1B2B2 homozygote. Thus, for disomic polyploids, self- fertilization can serve to prevent the disruption of favorable allelic interactions. Note that this implies that dominance intra-locus interactions of diploid genomes are converted to epistatic inter- locus interactions in disomic polyploids.

Defining Centromeres Genetically

Centromeres were defined genetically with a special mutant stock in Arabidopsis called quartet. Plants homozygous for the quartet mutation do not produce normal pollen grains, instead they produce pollen as groups of four microgametophytes that represent one microgametic tetrad.

That is, one cluster of four microgametophytes represent the four products of a single meiosis. Each tetrad of pollen can be separated and used to pollinate an emasculated tester plant, producing four seeds. These four seeds then can be analyzed using tetrad analysis procedures originally developed for fungi. By considering pairs of loci, one can classify each tetrad as "parental ditype" (PD), "nonparental ditype" (NPD), and "tetratype" (TT), see figure from Copenhaver et al. (1998):

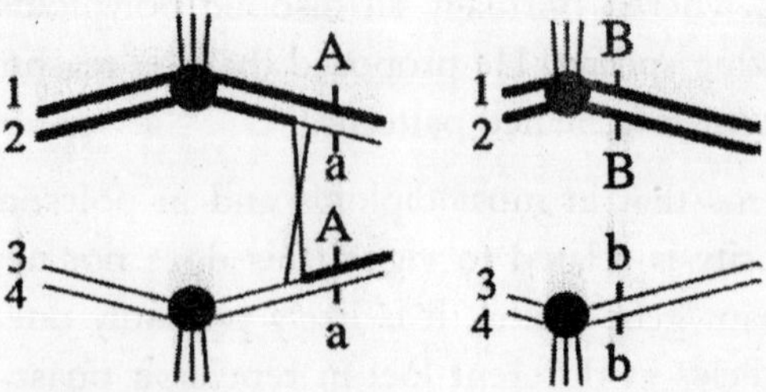

Tetrad member	PD	NPD	TT
i	A B	A b	A B
ii	A B	A b	a B
iii	a b	a B	A b
iv	a b	a B	a b

Copenhaver et al. (1998) screened 57 tetrads with 52 DNA markers, and later, Copenhaver et al. (1999) screened more than 1000 tetrads with DNA markers, including the centromere-associated 180 bp repeat. By identifying marker pairs *on different chromsomes* that showed few or no tetratype tetrads, they were able to genetically define the centromeres on the linkage maps defined by the genetic markers. To clarify their method, consider four loci on linkage group 1, two of which (A and B) are very near the centromere, such that crossovers between them and the centromere are rare, and the two of which (D and E) are far enough from the centromere that crossovers regularly occur between it and the centromere.

Then consider a second linkage group with loci F (near its centromere) and G (far from its centromere). Two parental lines with genotypes A1A1B1B1D1D1E1E1F1F1G1G1 and A2A2B2B2D2D2E2E2F2F2G2G2 are crossed to form the heterozygous genotype A1A2B1B2 D1D2E1E2F1F2G1G2, which also has the quartet mutation, so it forms pollen tetrads. This heterozygote is used to pollinate a tester genotype A3A3B3B3D3D3E3E3F3F3G3G3. The four testcross progeny corresponding to a pollen tetrad can be identified, permitting unordered tetrad analysis.

Assume that loci A and B flank the centromere so closely that crossovers very rarely or never occur between them and the centromere. Then consider the effect of a single crossover that occurs between loci B and D on linkage group 1 (C represents the centromere):

Chromatids at Meiosis I		Chromatids in resulting gametes
A1-C-B1————D1-E1		A1-C-B1————D1-E1
A1-C-B1————D1-E1		A1-C-B1————D2-E2
A2-C-B2————D2-E2	→	A2-C-B2————D1-E1
A2-C-B2————D2-E2		A2-C-B2————D2-E2

After this pollen tetrad is used to fertilize the tester stock, four progeny result. Ignoring the tester alleles, and considering the A and B loci, the progeny represent a Parental Ditype tetrad:

A1B1

A1B1

A2B2

A2B2

Ignoring the tester alleles, and considering the A and D loci, the result is a Tetratype:

A1D1

A1D2

A2D1

A2D2

The latter result also applies to the A and E pair, the B and D pair, and the B and E pair:

A1E1	B1D1	B1E1
A1E2	B1D2	B1E2
A2E1	B2D1	B2E1
A2E2	B2D2	B2E2 (all are tetratypes)

Now consider the D and E loci; the result is a Parental Ditype:

D1E1

D2E2

D1E1

D2E2

Remember that these tetrads are unordered, so the ditype shown above cannot be distinguished from the following order:

D1E1

D1E1

D2E2

D2E2

What happens if there are two crossovers between loci B and D? Non-parental ditypes for loci B and D can occur from a 4-strand double crossover that occurs between loci B and D, but this result could have also been due to a crossover that occurred between the centromere and locus A. A 2-strand double crossovers in the B-D interval will result in parental ditypes for B and D (indistinguishable from a zero-crossover meiosis), and 3-strand double crossovers will results in tetratypes for B and D.

Considering these results, you can see that *by comparing pairs of loci on the same chromosome*, you cannot distinguish between a low frequency of tetratypes that occurs because the two loci are both near the centromere or because the two loci are simply tightly linked and not near the centromere.

But, what happens if you consider pairs of loci that are on different chromosomes? Since the markers were already used to create a complete linkage map, you can know if the marker loci are on the same chromosome or not. Consider the case where there is a crossover between B and D on chromosome 1 and between F and G on chromosome 2.

The following tetrads could result:

B1F1	B1F2	
B1F1	B1F2	
B2F2	B2F1	
B2F2	B2F1 (no tetratypes possible)	
PD	NPD	
D1F1	D1F2	D1F1
D1F1	D1F2	D1F2
D2F2	D2F1	D2F1
D2F2	D2F1	D2F2
PD	NPD	TT
B1G1	B1G2	B1G1
B1G1	B1G2	B1G2
B2G2	B2G1	B2G1
B2G2	B2G1	B2G2

PD	NPD	TT
D1G1	D1G2	D1G1
D1G1	D1G2	D1G2
D2G2	D2G1	D2G1
D2G2	D2G1	D2G2
PD	NPD	TT

So, *when you look at two loci on different chromosomes and find no tetratypes, you know that they are both tightly linked to their cetromeres,* because the two loci themselves are not linked to each other.

Thus, Figure above of Copenhaver et al. (1998) illustrates the frequency of tetratypes between loci near the centromere on chromosomes I through IV on the one hand and all of the loci on chromosome V. The four lines correspond to the four centromere-associated loci on the other chromosomes, and they all indicate that their proportion of tetrads in pairs with chromosome V are minimum around position 74 cM, indicating that is the position of the centromere on chromosome V:

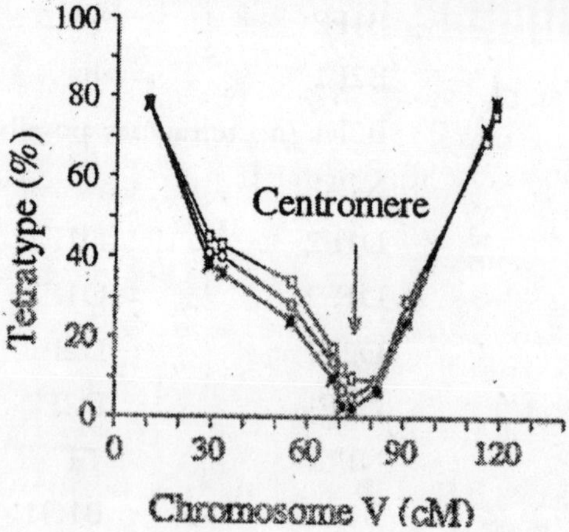

2

PLANT TISSUE CULTURE

INTRODUCTION

Most methods of plant transformation applied to GM crops require that a whole plant is regenerated from isolated plant cells or tissue which have been genetically transformed. This regeneration is conducted *in vitro* so that the environment and growth medium can be manipulated to ensure a high frequency of regeneration. In addition to a high frequency of regeneration, the regenerable cells must be accessible to gene transfer by whatever technique is chosen. The primary aim is therefore to produce, as easily and as quickly as possible, a large number of regenerable cells that are accessible to gene transfer. The subsequent regeneration step is often the most difficult step in plant transformation studies. However, it is important to remember that a high frequency of regeneration does not necessarily correlate with high transformation efficiency.

PLANT TISSUE CULTURE

Practically any plant transformation experiment relies at some point on tissue culture. There are some exceptions to this generalization, but the ability to regenerate plants from isolated cells or tissues *in vitro* underpins most plant transformation systems.

Plasticity and Totipotency

Two concepts, plasticity and totipotency, are central to understanding plant cell culture and regeneration. Plants, due to their long life span, have developed a greater ability to endure extreme conditions and predation than have animals. Many of the processes involved in plant growth and development adapt to environmental conditions. This plasticity allows plants to alter their metabolism, growth and development to best suit their environment. Particularly important aspects of this adaptation, as far as plant tissue culture and regeneration are concerned, are the abilities to initiate cell division from almost any tissue of the plant and to regenerate lost organs or undergo different developmental pathways in response to particular stimuli. When plant cells and tissues are cultured *in vitro* they generally exhibit a very high degree of plasticity, which allows one type of tissue or organ to be initiated from another type. In this way, whole plants can be subsequently regenerated. This regeneration of whole organisms depends upon the concept that all plant cells can, given the correct stimuli, express the total genetic potential of the parent plant. This maintenance of genetic potential is called 'totipotency'. Plant-cell culture and regeneration do, in fact, provide the most compelling evidence for totipotency.

The Culture Environment

When cultured *in vitro*, all the needs, both chemical and physical, of the plant cells have to met by the culture vessel, the growth medium and the external environment (light, temperature, etc.). The growth medium has to supply all the essential mineral ions required for growth and development. In many cases (as the biosynthetic capability of cells cultured *in vitro* may not replicate that of the parent plant), it must also supply additional organic supplements such as amino acids and vitamins.

Many plant cell cultures, as they are not photosynthetic, also require the addition of a fixed carbon source in the form of a sugar (most often sucrose). One other vital component that must also be supplied is water, the principal biological solvent. Physical factors, such as temperature, pH, the gaseous environment, light (quality and duration) and osmotic pressure, also have to be maintained within acceptable limits.

Plant Cell Culture Media

Culture media used for the *in vitro* cultivation of plant cells are composed of three basic components:

(1) essential elements, or mineral ions, supplied as a complex mixture of salts;

(2) an organic supplement supplying vitamins and/or amino acids; and

(3) a source of fixed carbon; usually supplied as the sugar sucrose.

For practical purposes, the essential elements are further divided into the following categories:

(1) macroelements (or macronutrients);

(2) microelements (or micronutrients); and

(3) an iron source.

Table 1. Some of the elements important for plant nutrition and their physiological function. These elements have to supplied by the culture medium in order to support the growth of healthy cultures *in vitro*

Element	Function
Nitrogen	Component of proteins, nucleic acids and some coenzymes Element required in greatest amount
Potassium	Regulates osmotic potential, principal inorganic cation
Calcium	Cell wall synthesis, membrane function, cell signalling
Magnesium	Enzyme cofactor, component of chlorophyll
Phosphorus	Component of nucleic acids, energy transfer, component of intermediates in respiration and photosynthesis
Sulphur	Component of some amino acids (methionine, cysteine) and some cofactors
Chlorine	Required for photosynthesis
Iron	Electron transfer as a component of cytochromes
Manganese	Enzyme cofactor
Cobalt	Component of some vitamins
Copper	Enzyme cofactor, electron-transfer reactions
Zinc	Enzyme cofactor, chlorophyll biosynthesis
Molybdenum	Enzyme cofactor, component of nitrate reductase

Complete, plant cell culture medium is usually made by combining several different components, as outlined in Table 2.

Media Components

It is useful to briefly consider some of the individual components of the stock solutions.

Macroelements

As is implied by the name, the stock solution supplies those elements required in large amounts for plant growth and development. Nitrogen, phosphorus, potassium, magnesium, calcium and sulphur (and carbon, which is added separately) are usually regarded as macroelements. These elements usually comprise at least 0.1% of the dry weight of plants.

Nitrogen is most commonly supplied as a mixture of nitrate ions (from the KNO_3) and ammonium ions (from the NH_4NO_3). Theoretically, there is an advantage in supplying nitrogen in the form of ammonium ions, as nitrogen must be in the reduced form to be incorporated into macromolecules. Nitrate ions therefore need to be reduced before incorporation.

However, at high concentrations, ammonium ions can be toxic to plant cell cultures and uptake of ammonium ions from the medium causes acidification of the medium. In order to use ammonium ions as the sole nitrogen source, the medium needs to be buffered. High concentrations of ammonium ions can also cause culture problems by increasing the frequency of vitrification (the culture appears pale and 'glassy' and is usually unsuitable for further culture). Using a mixture of nitrate and ammonium ions has the advantage of weakly buffering the medium as the uptake of nitrate ions causes OH^- ions to be excreted.

Phosphorus is usually supplied as the phosphate ion of ammonium, sodium or potassium salts. High concentrations of phosphate can lead to the precipitation of medium elements as insoluble phosphates.

Microelements

These elements are required in trace amounts for plant growth and development, and have many and diverse roles. Manganese, iodine, copper, cobalt, boron, molybdenum, iron and zinc usually comprise the microelements, although other elements such as nickel and aluminium are frequently found in some formulations.

Iron is usually added as iron sulphate, although iron citrate can also be used. Ethylene Diamine Tetra acetic Acid (EDTA) is usually used in conjunction with the iron sulphate. The EDTA complexes with the iron so as to allow the slow and continuous release of iron into the medium. Uncomplexed iron can precipitate out of the medium as ferric oxide.

Organic Supplements

Only two vitamins, thiamine (vitamin B1) and myo-inositol (considered a B vitamin) are considered essential for the culture of plant cells *in vitro*. However, other vitamins are also added to plant cell culture media.

Amino acids are also commonly included in the organic supplement. The most frequently used is glycine (arginine, asparagine, aspartic acid, alanine, glutamic acid, glutamine and proline are also used), but in many cases its inclusion is not essential.

Amino acids provide a source of reduced nitrogen and, like ammonium ions, uptake causes acidification of the medium. Casein hydrolysate can be used as a relatively cheap source of a mix of amino acids.

Carbon Source

Sucrose is cheap, easily available, readily assimilated and relatively stable and is therefore the most commonly used carbon source. Other carbohydrates (such as glucose, maltose, galactose and sorbitol) can also be used, and in specialized circumstances may prove superior to sucrose.

Table 2. Composition of a typical plant culture medium. The medium described here is that of Murashige and Skoog (MS)[a]

Essential element	Concentration in stock solution (mg/l)	Concentration in medium (mg/l)
Macroelements[b]		
NH_4NO_3	33000	1650
KNO_3	38000	1900
$CaCl_2.2H_2O$	8800	440
$MgSO_4.7H_2O$	7400	370
KH_2PO_4	3400	170
Microelements[c]		
KI	166	0.83
H_3BO_3	1240	6.2
$MnSO_4.4H_2O$	4460	22.3
$ZnSO_4.7H_2O$	1720	8.6
$Na_2MoO_4.2H_2O$	50	0.25
$CuSO_4.5H_2O$	5	0.025
$CoCl_2.6H_2O$	5	0.025
Iron source[c]		
$FeSO_4.7H_2O$	5560	27.8
$Na_2EDTA.2H_2O$	7460	37.3
Organic supplement[c]		
Myoinositol	20000	100
Nicotinic acid	100	0.5
Pyridoxine-HCl	100	0.5
Thiamine-HCl	100	0.5
Glycine	400	2
Carbon source[d]		
Sucrose	Added as solid	30 000

[a] Many other commonly used plant culture media (such as Gamborg's B5 and Schenk and Hildebrandt (SH) medium) are similar in composition to MS medium and can be thought of as 'high-salt' media. MS is an extremely widely used medium and forms the basis for many other media formulations.
[b] 50 ml of stock solution used per litre of medium.
[c] 5 ml of stock solution used per litre of medium.
[d] Added as solid.

Gelling Agents

Media for plant cell culture *in vitro* can be used in either liquid or 'solid' forms, depending on the type of culture being grown. For any culture types that require the plant cells or tissues to be grown on the surface of the medium, it must be solidified (more correctly termed 'gelled').

Agar, produced from seaweed, is the most common type of gelling agent, and is ideal for routine applications. However, because it is a natural product, the agar quality can vary from supplier to supplier and from batch to batch. For more demanding applications, a range of purer (and in some cases, considerably more expensive) gelling agents are available.

SUMMARY

These components, then, are the basic 'chemical' necessities for plant cell culture media. However, other additions are made in order to manipulate the pattern of growth and development of the plant cell culture.

PLANT GROWTH REGULATORS

We have already briefly considered the concepts of plasticity and totipotency. The essential point as far as plant cell culture is concerned is that, due to this plasticity and totipotency, specific media manipulations can be used to direct the development of plant cells in culture.

Plant growth regulators are the critical media components in determining the developmental pathway of the plant cells. The plant growth regulators used most commonly are plant hormones or their synthetic analogues.

Classes of Plant Growth Regulators

There are five main classes of plant growth regulator used in plant cell culture, namely:

(1) auxins;

(2) cytokinins;

(3) gibberellins;

(4) abscisic acid;

(5) ethylene.

Each class of plant growth regulator will be briefly looked at.

Auxins

Auxins promote both cell division and cell growth The most important naturally occurring auxin is IAA (indole-3-acetic acid), but its use in plant cell culture media is limited because it is unstable to both heat and light. Occasionally, amino acid conjugates of IAA (such as indole-acetyl-L-alanine and indole-acetyl-L-glycine), which are more stable, are used to partially alleviate the problems associated with the use of IAA. It is more common, though, to use stable chemical analogues of IAA as a source of auxin in plant cell culture media. 2,4-Dichlorophenoxyacetic acid (2,4-D) is the most commonly used auxin and is extremely effective in most circumstances. Other auxins are available, and some may be more effective or 'potent' than 2,4-D in some instances.

Table 3. Commonly used auxins, their abbreviation and chemical name

Abbreviation/name	Chemical name
2,4-D	2,4-dichlorophenoxyacetic acid
2,4,5-T	2,4,5-trichlorophenoxyacetic acid
Dicamba	2-methoxy-3,6-dichlorobenzoic acid
IAA	Indole-3-acetic acid
IBA	Indole-3-butyric acid
MCPA	2-methyl-4-chlorophenoxyacetic acid
NAA	1-naphthylacetic acid
NOA	2-naphthyloxyacetic acid
Picloram	4-amino-2,5,6-trichloropicolinic acid

CYTOKININS

Cytokinins promote cell division. Naturally occurring cytokinins are a large group of structurally related (they are purine derivatives) compounds. Of the naturally occurring cytokinins, two have some use in plant tissue culture media (see Table 4). These are zeatin and 2iP (2-isopentyl adenine). Their use is not widespread as they are expensive (particularly zeatin) and relatively unstable. The synthetic analogues, kinetin and BAP (benzylaminopurine), are therefore used more frequently. Non-purine-based chemicals, such as substituted phenylureas, are also used as cytokinins in plant cell culture media. These substituted phenylureas can also substitute for auxin in some culture systems.

Table 4. Commonly used cytokinins, their abbreviation and chemical name

Abbreviation/name	Chemical name
BAP[a]	6-benzylaminopurine
2iP (IPA)[b]	[N^6-(2-isopentyl)adenine]
Kinetin[a]	6-furfurylaminopurine
Thidiazuron[c]	1-phenyl-3-(1,2,3-thiadiazol-5-yl)urea
Zeatin[b]	4-hydroxy-3-methyl-trans-2-butenylaminopurine

[a] Synthetic analogues.
[b] Naturally occurring cytokinins.
[c] A substituted phenylurea-type cytokinin.

Gibberellins

There are numerous, naturally occurring, structurally related compounds termed 'gibberellins'. They are involved in regulating cell elongation, and are agronomically important in determining plant height and fruit-set. Only a few of the gibberellins are used in plant tissue culture media, Gibberelic Acid 3 (GA3) being the most common.

Abscisic Acid

Abscisic acid (ABA) inhibits cell division. It is most commonly used in plant tissue culture to promote distinct developmental pathways such as somatic embryogenesis.

Ethylene

Ethylene is a gaseous, naturally occurring, plant growth regulator most commonly associated with controlling fruit ripening, and its use in plant tissue culture is not widespread. It does, though, present a particular problem for plant tissue culture. Some plant cell cultures produce ethylene, which, if it builds up sufficiently, can inhibit the growth and development of the culture. The type of culture vessel used and its means of closure affect the gaseous exchange between the culture vessel and the outside atmosphere and thus the levels of ethylene present in the culture.

Plant Growth Regulators and Tissue Culture

Generalisations about plant growth regulators and their use in plant cell culture media have been developed from initial observations made in the 1950s. There is, however, some considerable difficulty in predicting the effects of plant growth regulators: this is because of the great differences in culture response between species, cultivars and even plants of the same cultivar grown under different conditions.

However, some principles do hold true and have become the paradigm on which most plant tissue culture regimes are based. Auxins and cytokinins are the most widely used plant growth regulators in plant tissue culture and are usually used together, the ratio of the auxin to the cytokinin determining the type of culture established or regenerated (see Figure 1). A high auxin to cytokinin ratio generally favours root formation, whereas a high cytokinin to auxin ratio favours shoot formation. An intermediate ratio favours callus production.

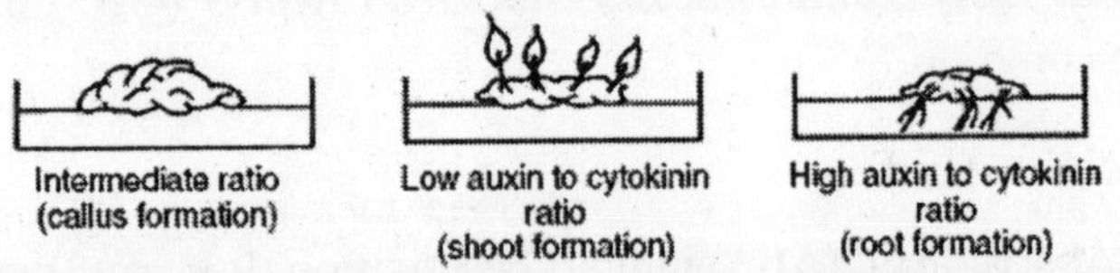

Figure 1. The effect of different ratios of auxin to cytokinin on the growth and morphogenesis of callus. High auxin to cytokinin ratios promote root development, low ratios promote shoot development. Intermediate ratios promote continued growth of the callus without differentiation.

CULTURE TYPES

Cultures are generally initiated from sterile pieces of a whole plant. These pieces are termed 'explants', and may consist of pieces of organs, such as leaves or roots, or may be specific cell types, such as pollen or endosperm. Many features of the explant are known to affect the efficiency of culture initiation. Generally, younger, more rapidly growing tissue (or tissue at an early stage of development) is most effective. Several different culture types most commonly used in plant transformation studies will now be examined in bit detail.

Callus

Explants, when cultured on the appropriate medium, usually with both an auxin and a cytokinin, can give rise to an unorganised, growing and dividing mass of cells. It is thought that any plant tissue can be used as an explant, if the correct conditions are found. In culture, this proliferation can be maintained more or less indefinitely, provided that the callus is subcultured on to fresh medium periodically.

During callus formation there is some degree of dedifferentiation (i.e. the changes that occur during development and specialization are, to some extent, reversed), both in morphology (callus is usually composed of unspecialised parenchyma cells) and metabolism. One major consequence of this dedifferentiation is that most plant cultures lose the ability to photosynthesise. This has important consequences for the culture of callus tissue, as the metabolic profile will probably not match that of the donor plant. This necessitates the addition of other components—such as vitamins and, most importantly, a carbon source—to the culture medium, in addition to the usual mineral nutrients.

Callus culture is often performed in the dark (the lack of photosynthetic capability being no drawback) as light can encourage differentiation of the callus. During long-term culture, the culture may lose the requirement for auxin and/or cytokinin. This process, known as 'habituation', is common in callus cultures from some plant species (such as sugar beet). Callus cultures are extremely important in plant biotechnology. Manipulation of the auxin to cytokinin ratio in the medium can lead to the development of shoots, roots or somatic embryos from which whole plants can subsequently be produced. Callus cultures can also be used to initiate cell suspensions, which are used in a variety of ways in plant transformation studies.

CELL-SUSPENSION CULTURES

Callus cultures, broadly speaking, fall into one of two categories: compact or friable. In compact callus the cells are densely aggregated, whereas in friable callus the cells are only loosely associated with each other and the callus becomes soft and breaks apart easily. Friable callus provides the inoculum to form cell-suspension cultures. Explants from some plant species or particular cell types tend not to form friable callus, making cell-suspension initiation a difficult task. The friability of callus can sometimes be improved by manipulating the medium components or by repeated subculturing. The friability of the callus can also sometimes be improved by culturing it on 'semi-solid' medium (medium with a low concentration of gelling agent). When friable callus is placed into a liquid medium (usually the same composition as the solid medium used for the callus culture) and then agitated, single cells and/or small clumps of cells are released into the medium.

Under the correct conditions, these released cells continue to grow and divide, eventually producing a cell-suspension culture. A relatively large inoculum should be used when initiating cell suspensions so that the released cell numbers build up quickly. The inoculum should not be too large though, as toxic products released from damaged or stressed cells can build up to lethal levels. Large cell clumps can be removed during subculture of the cell suspension. Cell suspensions can be maintained relatively simply as batch cultures in conical flasks. They are continually cultured by repeated subculturing into fresh medium. This results in dilution of the suspension and the initiation of another batch growth cycle. The degree of dilution during subculture should be determined empirically for each culture. Too great a degree of dilution will result in a greatly extended lag period or, in extreme cases, death of the transferred cells.

After subculture, the cells divide and the biomass of the culture increases in a characteristic fashion, until nutrients in the medium are exhausted and/or toxic by-products build up to inhibitory levels—

this is called the 'stationary phase'. If cells are left in the stationary phase for too long, they will die and the culture will be lost. Therefore, cells should be transferred as they enter the stationary phase. It is therefore important that the batch growth-cycle parameters are determined for each cell-suspension culture.

Protoplasts

Protoplasts are plant cells with the cell wall removed. Protoplasts are most commonly isolated from either leaf mesophyll cells or cell suspensions, although other sources can be used to advantage. Two general approaches to removing the cell wall (a difficult task without damaging the protoplast) can be taken—mechanical or enzymatic isolation. Mechanical isolation, although possible, often results in low yields, poor quality and poor performance in culture due to substances released from damaged cells. Enzymatic isolation is usually carried out in a simple salt solution with a high osmoticum, plus the cell wall degrading enzymes. It is usual to use a mix of both cellulase and pectinase enzymes, which must be of high quality and purity. Protoplasts are fragile and easily damaged, and therefore must be cultured carefully. Liquid medium is not agitated and a high osmotic potential is maintained, at least in the initial stages. The liquid medium must be shallow enough to allow aeration in the absence of agitation. Protoplasts can be plated out on to solid medium and callus produced. Whole plants can be regenerated by organogenesis or somatic embryogenesis from this callus. Protoplasts are ideal targets for transformation by a variety of means.

ROOT CULTURES

Root cultures can be established *in vitro* from explants of the root tip of either primary or lateral roots and can be cultured on fairly simple media. The growth of roots *in vitro* is potentially unlimited, as roots are indeterminate organs. Although the establishment of root cultures was one of the first achievements of modern plant tissue culture, they are not widely used in plant transformation studies.

Shoot Tip and Meristem Culture

The tips of shoots (which contain the shoot apical meristem) can be cultured *in vitro*, producing clumps of shoots from either axillary or adventitious buds. This method can be used for clonal propagation. Shoot meristem cultures are potential alternatives to the more commonly used methods for cereal regeneration (see the Case study below) as they are less genotype-dependent and more efficient (seedlings can be used as donor material).

Embryo Culture

Embryos can be used as explants to generate callus cultures or somatic embryos. Both immature and mature embryos can be used as explants. Immature, embryo-derived embryogenic callus is the most popular method of monocot plant regeneration.

Microspore Culture

Haploid tissue can be cultured *in vitro* by using pollen or anthers as an explant. Pollen contains the male gametophyte, which is termed the 'microspore'. Both callus and embryos can be produced from pollen. Two main approaches can be taken to produce *in vitro* cultures from haploid tissue.

The first method depends on using the anther as the explant. Anthers (somatic tissue that surrounds and contains the pollen) can be cultured on solid medium (agar should not be used to solidify the medium as it containsinhibitory substances). Pollen-derived embryos are subsequently produced via dehiscence of the mature anthers. The dehiscence of the anther depends both on its isolation at the correct stage and on the correct culture conditions. In some species, the reliance on natural dehiscence can be circumvented by cutting the wall of the anther, although this does, of course, take a considerable amount of time. Anthers can also be cultured in liquid medium, and pollen released from the anthers can be induced to form embryos, although the effi-ciency of plant regeneration is often very low. Immature pollen can also be extracted from developing anthers and cultured directly, although this is a very time-consuming process.

Both methods have advantages and disadvantages. Some beneficial effects to the culture are observed when anthers are used as the explant material. There is, however, the danger that some of the embryos produced from anther culture will originate from the somatic anther tissue rather than the haploid microspore cells. If isolated pollen is used there is no danger of mixed embryo formation, but the efficiency is low and the process is time-consuming. In microspore culture, the condition of the donor plant is of critical importance, as is the timing of isolation. Pretreatments, such as a cold treatment, are often found to increase the efficiency. These pretreatments can be applied before culture, or, in some species, after placing the anthers in culture.

Plant species can be divided into two groups, depending on whether they require the addition of plant growth regulators to the medium for pollen/anther culture; those that do also often require organic supplements, e.g. amino acids. Many of the cereals (rice, wheat, barley and maize) require medium supplemented with plant growth regulators for pollen/anther culture.

Regeneration from microspore explants can be obtained by direct embryogenesis, or via a callus stage and subsequent embryogenesis.

Haploid tissue cultures can also be initiated from the female gametophyte (the ovule). In some cases, this is a more efficient method than using pollen or anthers.

The ploidy of the plants obtained from haploid cultures may not be haploid. This can be a consequence of chromosome doubling during the culture period. Chromosome doubling (which often has to be induced by treatment with chemicals such as colchicine) may be an advantage, as in many cases haploid plants are not the desired outcome of regeneration from haploid tissues. Such plants are often referred to as 'di-haploids', because they contain two copies of the same haploid genome.

PLANT REGENERATION

Having looked at the main types of plant culture that can be established *in vitro*, we can now look at how whole plants can be regenerated from these cultures.

In broad terms, two methods of plant regeneration are widely used in plant transformation studies, i.e. somatic embryogenesis and organogenesis.

Somatic Embryogenesis

In somatic (asexual) embryogenesis, embryo-like structures, which can develop into whole plants in a way analogous to zygotic embryos, are formed from somatic tissues (Figure 2). These somatic embryos can be produced either directly or indirectly. In direct somatic embryogenesis, the embryo is formed directly from a cell or small group of cells without the production of an intervening callus. Though common from some tissues (usually reproductive tissues such as the nucellus, styles or pollen), direct somatic embryogenesis is generally rare in comparison with indirect somatic embryogenesis.

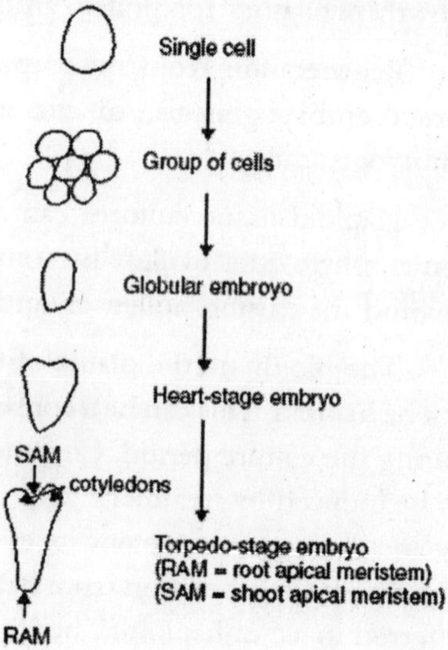

Figure 2. A schematic representation of the sequential stages of somatic embryo development. Somatic embryos may develop from single cells or from a small group of cells. Repeated cell divisions lead to the production of a group of cells that develop into an organised structure known as a 'globular-stage embryo'. Further development results in heart- and torpedo-stage embryos, from which plants can be regenerated. Zygotic embryos undergo a fundamentally similar development through the globular (which is formed after the 16-cell stage), heart and torpedo stages. Polarity is established early in embryo development. Signs of tissue differentiation become apparent at the globular stage and apical meristems are apparent in heart-stage embryos.

In indirect somatic embryogenesis, callus is first produced from the explant. Embryos can then be produced from the callus tissue or from a cell suspension produced from that callus.

Somatic embryogenesis usually proceeds in two distinct stages. In the initial stage (embryo initiation), a high concentration of 2,4-D is used. In the second stage (embryo production) embryos are produced in a medium with no or very low levels of 2,4-D.

In many systems it has been found that somatic embryogenesis is improved by supplying a source of reduced nitrogen, such as specific amino acids or casein hydrolysate.

Organogenesis

Somatic embryogenesis relies on plant regeneration through a process analogous to zygotic embryo germination.

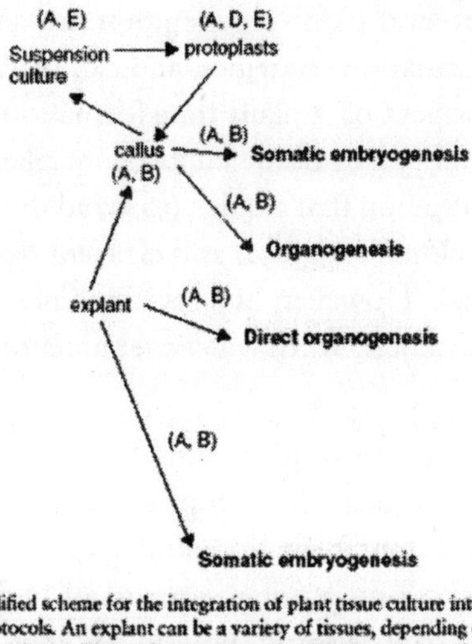

Figure 3. A simplified scheme for the integration of plant tissue culture into plant transformation protocols. An explant can be a variety of tissues, depending on the particular plant species being cultured. The explant can be used to initiate a variety of culture types, depending on the explant used. Regeneration by either organogenesis or somatic embryogenesis results in the production of whole plants. Different culture types and regeneration methods are amenable to different transformation protocols. The transformation protocols (see Chapter 3) highlighted in this figure are: (A) *Agrobacterium*-mediated; (B) biolistic transformation; (D) direct DNA uptake and (E) electroporation. Different combinations of culture type and transformation protocol are used depending on the plant species and cultivar being used. In some species a variety of culture types and regeneration methods can be used, which enables a wide variety of transformation protocols to be utilised. In other species there is effectively no choice over culture type and/or regeneration method, which can limit the transformation protocols that are applicable. (Redrawn with permission from Walden R. and Wingender R. (1995).)

Organogenesis relies on the production of organs, either directly from an explant or from a callus culture. There are three methods of plant regeneration via organogenesis. The first two methods depend on adventitious organs arising either from a callus culture or directly from an explant (Figure 3). Alternatively, axillary bud formation and growth can also be used to regenerate whole plants from some types of tissue culture. Organogenesis relies on the inherent plasticity of plant tissues, and is regulated by altering the components of the medium. In particular, it is the auxin to cytokinin ratio of the medium that determines which developmental pathway the regenerating tissue will take. It is usual to induce shoot formation by increasing the cytokinin to auxin ratio of the culture medium. These shoots can then be rooted relatively simply.

SUMMARY

Tissue culture and plant regeneration are an integral part of most plant transformation strategies, and can often prove to be the most challenging aspect of a plant transformation protocol. Key to success in integrating plant tissue culture into plant transformation strategies is the realisation that a quick (to avoid too many deleterious effects from somaclonal variation) and efficient regeneration system must be developed. However, this system must also allow high transformation efficiencies from whichever transformation technique is adopted.

Not all regeneration protocols are compatible with all transformation techniques. Some crops may be amenable to a variety of regeneration and transformation strategies, others may currently only be amenable to one particular protocol. Advances are being made all the time, so it is impossible to say that a particular crop will never be regenerated by a particular protocol. However, some protocols, at least at the moment, are clearly more efficient than others. Regeneration from immature embryo-derived somatic embryos is, for example, the favoured method for regenerating monocot species.

3

PLANT CYTOGENETICS: A RE-BIRTH IN 21ST CENTURY

During the last few decades of the last century, emphasis on classical plant cytogenetics largely declined due to the emergence of molecular biology tools. In the realm of plant cytogenetics, this marked the end Era 1 dealing with the study of structural and numerical changes of chromosomes, and also Era 2 dealing with DNA content and composition (repetitive vs unique DNA). In Era 3, during 1980s and 1990s, molecular markers were extensively used for construction of molecular linkage maps and physical maps of chromosomes in a variety of plants. Molecular markers were also used and for the study of marker-trait associations for marker-aided selection (MAS). Later, in Era 4, starting in mid 1990s, two major areas of research took over; these were plant genomics involving whole genome sequencing (sequencing of the genomes of Arabidopsis thaliana and Oryza sativa), and plant epigenomics involving study of nuclear architecture, chromatin remodeling (chromatin remodeling complexes, and histone modifications and variants; the histone code). Some details of chromosome organization within the nucleus, and genomics/ epigenomics research are discussed in this discussion.

INTRODUCTION

Soon after the rediscovery of Mendel's Laws in the year 1900, the formulation of 'chromosome theory of inheritance' laid the foundation of cytogenetics, which has thus completed its first hundred glorious years in the year 2003. Subsequently, in the last few decades of the last century, the interest in plant cytogenetics that involved study and use of structural and numerical changes of chromosomes had largely declined. However, more recently, with interest in construction of molecular maps of chromosomes, and in the sequencing of whole genomes comprising an entire set of chromosomes, plant cytogenetics perhaps had a re-birth.

Other areas of recent research in plant cytogenetics include study of the organization of individual chromosomes in the form of chromosome territories and location of replication and transcription territories within the nucleus.

Chromosome structure has also been studied in great detail, so that the chromatin is now known to consist of nucleosome and chromatosome subunits, which undergo several levels of folding to contain it within the boundaries of the chromosomes. The structure of these nucleosome subunits has been studied at atomic resolution, and the structural relationship between adjacent nucleosomes has been recently resolved through X-ray crystallography of tetanucleosomes (Schalch et al., 2005).

The structure of telomeres and centromeres have also been studied in several plant systems (particularly the cereals) in great detail suggesting that while structure of telomeres is conserved, but that of centromeres exhibit sufficient diversity.

In recent years, it has also been shown that both DNA and histone proteins, as components of chromosomes undergo large-scale modifications, which regulate gene expression. The study of these modifications also gave birth to the science of 'epigenetics and epigenomics'.

Projects like whole genome/epigenome sequencing of important plant systems are also yielding new information, giving new directions to research in plant cytogenetics. These exciting new developments in the field of plant cytogenetics will be briefly discussed in this discussion.

FOUR OVERLAPPING ERAS OF CYTOGENETICS RESEARCH

Progress of plant cytogenetics during the last ~80 years can be broadly divided in the following four eras, which may be slightly overlapping. Cytogenetics Era 1 (1910-1970): In this era chromosome number of a number of plant systems became known, structural changes like interachanges and inversions were studied for the first time in Stizolobium, Datura and Oentothera, inversions were studied in maize, and anuploids developed and cytogenetic maps constructed in several crops like maize, wheat, barley, etc. Alien addition and substitution lines were also developed in bread wheat using rye and few Aegilops/Agropyron species as the source of alien chromosomes.

Cytogenetics Era 2 (1950-1980): In this era, haploid DNA content (C-value) and composition (unique and repetitive) of nuclear DNA were determined in a large number of flowering plants using techniques of cytospectrophotometry and reassociation kinetics. This led to recognition of two versions of C-value paradox. Firstly, the DNA contents in most eukaryotes were too high for the number of genes in the corresponding taxa, as estimated on the basis of known rates of mutations, and secondly the large-scale variation in DNA contents, could not be explained with the level of difference in complexity witnessed in these different organisms. The occurrence of large proportion of repetitive DNA in each of these eukaryotic genomes partly resolved the C-value paradox.

In this era, starting in early 1980s, DNA-based molecular markers were developed and molecular maps constructed in a large number of animals and plant systems, so that these molecular markers and the corresponding maps became an important resource for a variety of research problems, including their use in diagnostics and plant breeding. During this period, another significant development was the availability of a variety of fluorescence in situ hybridization (FISH), including multicolour FISH (McFISH), chromosome orientation FISH (CO-FISH), fibre-FISH, RNA-FISH, Comparative Genomic Hybridization (CGH) and 3-D FISH.

This era started in mid-1990s and gained momentum in the present century, with two distinct areas of cytogenetics research; first, the whole genome sequencing giving birth to 'reverse genetics', and second, the chromatin remodeling giving birth to the concept of 'histone code'. It is thus obvious that chromosome research has evolved and progressed at an incredible pace in the last two decades, more particularly during the past ten years. Since details of research done during Era 1, Era 2 and Era 3 would be widely available in text books and review articles, we mainly discuss the progress made during Era 4, in which significant progress has been made during the last few years to elucidate how the nucleosome and chromatin structure are modulated for expression of genes in time and space.

NUCLEAR ARCHITECTURE AND CHROMOSOMES

Chromosome organization within the nucleus: In the past, it was believed that chromatin within the nucleus is a network, intertwined and randomly distributed within the space available in the nucleus. However, both in plants and animal systems, recent evidence has demonstrated that the nucleus is a highly compartmentalized structure. Chromosome territories and interchromatin compartment (CT-IC model): The chromatin within the nucleus is organized in the form of chromosome territories(CTs) and interchromatin compartments (IC), and hence the formulation of CT-IC model. While CTs contain individual chromosomes, IC contains macro molecular complexes that are needed for replication, transcription, splicing and repair.

Newer techniques combining 3-D-FISH and computer aided deconvolution techniques helped in resolving the following features of chromatin organization and behaviour

(i) in an interphase nucleus, each individual chromosome occupies a discrete space, called the 'chromosome territory' or CT and that there is little inter twining among chromosomes;

(ii) in interphase cells, each chromosome also interacts with the nuclear envelope through consistent contact points;

(iii) in interphase cells, each chromosome interacts with other chromosomes through its heterochromatic regions; and

(iv) in dividing cells, chromosome movements are also non-random. The above information regarding chromosome organization at the physical level has also been integrated with genetic and molecular data, to decipher the mechanisms of different nuclear processes, in which chromosomes are involved, more particularly the transcription and DNA replication.

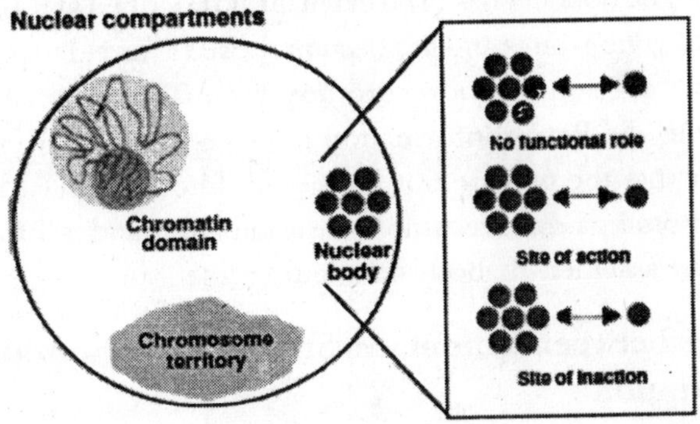

Figure. Compartmentalization in the mammalian nucleus. The nucleus contains proteinaceous nuclear bodies, chromatin domains including heterochromatin and euchromatin and chromosome territories. Nuclear bodies can either be non-specific aggregates, sites of nuclear processes (rRNA transcription in the nucleolus) or sites of inaction (storage of splicing components in splicing factor compartments).

Rabl organization and telomeres orientation: Chromosome segregation at anaphase results in the polarization of chromosomes because sister centromeres are pulled in opposite directions and the rest of the chromosome arms trail behind. In some instances, this anaphase arrangement of chromosomes persists into the following interphase; this is known as the Rabl organization, in which chromosomes have a preferentially polarized organization, with centromeres at one end of the nuclear envelope, called the apical side, and telomeres at the opposite end, called the basal side. The presence of the Rabl organization is known to vary greatly between species and among tissues or developmental stages of an organism. In plants, it is generally observed in species with bigger genomes like those of wheat, rye, barley, and oats, but not in species with smaller genomes like those of sorghum and rice. Maize with intermediate size of genome displayed neither entirely Rabl nor entirely random chromosome organization.

Bouquet and telomeres clustering: The bouquet is the clustering of chromosome ends on the nuclear envelope (NE) during meiotic prophase, coincident with the initiation of homologous chromosome synopsis. The bouquet has been extensively described in many species in all eukaryotic groups and has been proposed as an aid to presynaptic alignment of homologous chromosomes. The similarity of the bouquet to the Rabl conformation has long been noted. However, it is clear that the two are not the same, although similar functions (chromosome pairing, recombination initiation and SC formation) have been assigned to both of them.

Relation between Chromatin Structure/ organization and Transcription

Heterochromatin and euchromatin: We know that the chromosomes of eukaryotes consist of darkly stained heterochromatin and lightly stained euchromatin. According to the classical view, heterochromatin is transcriptionally inactive and euchromatin is active.

This view is changing now; following are some examples: (i) there is evidence that several repeats within heterochromatic centromeres are transcribed to produce siRNA; (ii) it has been show that in each of the five chromosomes of Arabidopsis, heterochromatin is largely confined to the pericentromeric regions, and mainly consists of 180-bp satellite repeats, and retrotransposons (mainly from Athila family);(iii) NORs, mainly consisting of repetitive DNA, carries thousands of kilobases of tandemly repeated ribosomal DNA encoding rRNA. It has now been conclusively proved that in eukaryotic chromatin, cytosine methylation in repetitive DNA and distinctive modifications of individual histone proteins are frequent and cause heterochromatin formation. There are at least three kinds of methylases causing cytosine methylation in DNA: (i) Dnmt-1 type DNA methylase adds methyl-residues to cytosines (CG or CNG?) on newly synthesized strand of a DNA duplex, by using information from the conserved parental DNA strand carrying met-C, thus making DNA methylation heritable; (ii) Dnmt-3 type DNA methylase brings about DNA methylation (CG or CNG?) de novo, so that unmethylated DNA becomes methylated on both strands; (iii) chromomethylases (unique to plants) in Arabidopsis and maize recognize CpNpG sequences and bring about cytosine methylation.

TRANSCRIPTION AND REPLICATION FACTORIES : A NEW CONCEPT

In recent years, specific regions have been identified within the eukaryotic nucleus, where replication and transcription takes place. It has also been shown that transcription sites are spatially distinct from replication sites. Therefore, we need to realize thattranscription by Pol II is not homogeneously distributed throughout the nucleoplasm, but occurs at highly enriched Pol II foci, known as transcription factories, which contain most of the hyperphosphorylated, elongating form of Pol II. Similarly replication takes place in 'replication factories'. It has been demonstrated that the chromosome segments destined to be transcribed or replicated have to move physically to these regions for transcription or replication to take place. Peter Fraser and his co-workers from Cambridge described this subject in some detail (Chakalova et al., 2005).

Transcription factories: A transcription factory is generally 80 nm tripartate structure mainly containing three spatially contiguous regions, with the template, the RNA polymerase II (Pol II) and the newly synthesised mRNA. The Pol II is perhaps attached to the transcription site, so that the template would move along RNAP II rather than the Pol II tracking along the template. Some translation also appears to be coupled with transcription at transcription factories.

There are fewer transcription factories than there are active genes and other transcription units in the nucleus, so that that more than one active gene is transcribed in each factory. And actively transcribed genes that are separated by long distances frequently co-localize in the same transcription factory. It has also been shown that actively transcribed genes co-localize with transcription factories, whereas identical, temporarily non-transcribed alleles, which can often be in the same cell, do not. Therefore, the 'on' state correlates with factory occupancy and the 'off' state with relocation away from factories. The specific Thus the specific nuclear repositioning of genes is correlated with transcriptional activation, silencing and replication timing.

The fact that different genes frequently co-occupy the same factory provides strong evidence that genes do not assemble their own transcription sites de novo when they become active, but instead migrate to preassembled transcription sites. A stable factory implies that genes or transcription units would essentially be pulled through a factory, rather than polymerases moving along the chromatin fibre, as is commonly believed. The finding that approximately 15% of the genome is transcribed — although probably not all at once — indicates that an extraordinarily large part of the genome passes through the limited number of transcription factories in a cell nucleus. This must have a profound effect on the nuclear organization of the genome.

Replication factories: It has been shown that a segment of DNA that needs to be replicated temporarily disengages Pol II and ceases transcription, although transcription in other regions of the genome continues uninterrupted throughout S phase. This segment of DNA needs to relocate itself in a replication factory. There seems to be no overlap between replication and transcription sites/factories.

Chromatin Remodeling

The term "chromatin-remodeling" generally refers to changes in histone-DNA interactions in nucleosomes. Histone proteins are modified and non-histone chromosomal proteins (e.g. high mobility group nuclear proteins = HMGN proteins; heterochromatin protein = HP1) are indirectly involved in the modification and activity of chromatin. Histone-DNA interactions in the nucleosomes are also modulated (remodeled) to facilitate interaction of other factors (not involved in remodeling) with DNA template. It is believed that the histone octamers of nucleosomes are often displaced from the enhancer and promoter regions by chromatin remodeling complexes to allow access of a variety of factors to DNA. During 1990s, evidence for chromatin remodeling and cellular memory became available in yeast (Saccharomyces cerevisiae), fruitfly (Drosophila melanogaster), and mammals, through genetic and biochemical studies.However, later towards the end of the last century and in the early years of the present century, chromatin remodeling has been studied and factors associated with this phenomenon in plant systems like Arabidopsis.

Activities catalyzed by chromatin remodeling: Chromatin-remodeling factors can catalyze the following activities:

(i) mobilization and repositioning of nucleosomes,

(ii) transfer of a histone octamer from a nucleosome to a separate DNA template,

(iii) the facilitated access of nucleases (enzymes) to nucleosomal DNA,

(iv) creation of di-nucleosome-like structures from mono-nucleosomes,

(v) generation of superhelical torsion in DNA, (vi) disruption of histone-DNA contacts; and

(vi) assembly and disassembly of nucleosomes.

Components of chromatin remodeling complexes: Three major strategies are used for chromatin remodeling. Each strategy makes use of a different set of proteins, described as components of chromatin remodeling complexes. Following are the three classes of chromosome remodeling complexes/components:

(i) ATP-dependent chromatin remodeling complex, which makes use of ATP for chromatin modification through mere histone-DNA interactions.

(ii) Histone modifying enzymes, which are used for post-translational modifications of histones (mainly acetylation and methylation); these modifications create signals that define the so-called 'histone code'.

(iii) Variants of the histones (H2A, H2B, H3, H4 and H), which are synthesized and incorporated into nucleosome subunits (in place of normal histones), bringing about chromatin remodeling. More recently, the extreme use of H3 histone variants also led to the formulation of 'H3 barcode hypothesis' (Hake and Allis, 2006).

ATP-dependent chromatin remodeling machines/complexes: ATP-dependent :chromatin remodeling complexes (CRCs); also called chromosome remodeling machines (CRM), consist of ATPase polypeptide subunits and also non-ATPase subunits. Although ATPases play an major important role in chromatin remodeling, non-ATPase components also have a role to play. The ATPases can be quite diverse and not all ATPases are important for chromosome remodeling (see below). These ATPases are classified into three superfamilies, each superfamily having several families. One of these families belongs to a specific class of ATPases, described as 'Snf2-like family of ATPases', which are relevant to chromosome remodeling and therefore will be discussed in some detail.

(a) ATPases as subunits of ATP- dependent complexes: Members of SNF2-like family of ATPases are classified in several subfamilies, depending upon which protein motifs outside the ATPase region they have. Atleast seven subfamilies are known, but only four subfamilies are important, which include the following:

(i) Swi/Snf2 subunit of the SWI/SNF complex;

(ii) ISWI (ISWI, hSNF2H, hSNF2L, yISW1, yISW2; h stands for human, and y stands for yeast);

(iii) CHD1 (CHD1, Mi-2a/CHD3, Mi-2b/CJD4, Hrp1, Hrp3),

(iv) INO80 (CSB, Rad.26, ERCC6).

(b) Non-ATPase subunits of ATP- dependent complexes: The ATP dependent chromatin remodeling complexes also have non-ATPase subunits, which may have the following functions:

(i) enhance or regulate motor activity of ATPase subunits;

(ii) mediate other specialized functions that are not related with chromatin remodeling– these may be recruited to promoters via interactions with sequence specific transcription factors (TFs). Rad54 is also an example, in which SNF2-like ATPase is programmed by another polypeptide. Rad54 and Rad51 (related with bacterial RecA protein) catalyze homologous strand pairing.

Chromosome remodeling through histone modifications: The 'histone code': In eukaryotes the fundamental unit of chromatin is the nucleosome, which is a protein octamer/DNA complex composed of 200 bp wrapped around a histone octamer that consists of two molecules each of four core histones,

H2A, H2B, H3 and H4. One molecule of linker histone H1 is also associated with each nucleosome. Crystal structure of the core particle of this nucleosome unit consisting of 146 bp of DNA wrapped around histone octamer was determined in late 1990s at 2.8 A resolution (Luger et al., 1997).

More recently, crystal structure of a tetranucleosome was also determined at 9 Å resolution to understand the manner of higher-order folding of the nucleosome sub-units (Schalch et al., 2005).

It was also shown in several recent studies that amino-terminal tails of histone proteins are targets for a series of post-translational modifications (PTMs), including acetylation, phosphorylation, and methylation. These modifications regulate chromatin structure and gene expression (Strahl and Allis, 2000; Jenuwein and Allis, 2001; Turner, 2002). Multiple histone modifications in various combinations are thought to form a 'histone code', since these modifications extend the information capacity of the associated DNA (Strahl and Allis, 2000).

For example, histone H3 and H4 acetylation is consistently associated with transcriptionally active euchromatin, while methylation can be associated with either active or inactive chromatin depending on the residue involved in methylation.

For instance, methylation at H3K4, H3K36 and H3K79 are hallmarks for active transcription, whereas methylation at H3K9, H3K27 and H4K20 are correlated with transcriptionally inert heterochromatin (Fischle et al., 2003; Lachner et al., 2003; Margueron et al., 2005; Peters and Schubeler, 2005). Further, the lysines in histone N-terminal tails can be mono-methylated (me1), di-methylated (me2) or tri-methylated (me3), and each methylation state may have unique biological functions, increasing the potential complexity of the histone code (Dutnall, 2003).

Table. Histone modification combinations and their effect on transcription

Nature of histone Modification	Effect on transcription
Histone acetylation	Transcription activation
Histone methylation	
H3K4, H3K36 and H3K79	Transcription activation
H3K9, H3K27 and H4K20	Transcription silencing

Table. Various combinations of histone methylations and their effect on state of chromatin condensation

Degree of methylation	State of chromatin
H3K9me3, H3K27me1 and H4K20me3	Highly condensed heterochromatin
H3K9me1, H3K9me2 and H3K27me3	Lightly condensed heterochromatin

It has been shown that methylation of H3K9, H3K27 and H4K20 is involved in heterochromatinization, but the degree of methylation of each of these lysine residues determines the degree of heterochromatinization (Peters et al., 2003; Schotta et al., 2004).

For instance, H3K9me3, H3K27me1 and H4K20me3 mark the most deeply stained regions while H3K9me1, H3K9me2 and H3K27me3 mark the less condensed heterochromatin (Peters et al., 2003; Plath et al., 2003; Rice et al., 2003; Silva et al., 2003; Schotta et al., 2004; Okamoto et al., 2004). Several reports indicate that the monoand di-methylated forms of H3K9 and H3K27 are enriched in heterochromatin (Jackson et al., 2004;

Lindroth et al., 2004; Mathieu et al., 2005; Naumann et al., 2005), although degree of methylation also depends on the genome size (Houben et al., 2003).

For instance, unlike in animals, in Arabidopsis, H3K27me3 is associated with euchromatin and H3K9me3 is extremely rare (Lindroth et al., 2004; Mathieu et al., 2005; Naumann et al., 2005).

There is also evidence that H4K20 methylation is also present in Arabidopsis (Naumann et al., 2005; NG et al., 2006). It is not clear, however, whether there is a direct relationship between heterochromatin and histone methylation. However, in plants with large genomes including maize and barley and wheat, very little is known about histone methylation (Houben et al., 2003), which makes up the bulk of the angiosperms (Arumuganthan and Earle, 1991).

In a recent study, quantitative distribution of mono-, di-, and trimethylation at H3K9 and H3K27 was examined in maize on whole genome level using oachytene chromosomes. The data reveal that three marks (H3K9me1, H3K27me1, and H3K27me2) correlate with DAPI (DNA) staining, but that only H3K27me2 is specifically enriched in condensed areas. It was also observed that H3K9me2 is not abundant in heterochromatin, but is instead enriched between chromomeres along with H3K4me2. The study also reported that centromeres contain H3K9me2 and H3K9me3; that H3K27me3 occurs at several brightly focused euchromatic domains, and that H4K20 methylation is rare or absent.

The Histone Code Hypothesis states that chromatin-DNA interactions are guided by combinations of histone modifications. For example, phosphorylation of serine residues 10 and 28 on H3 is a marker for chromosomal condensation; similarly, phosphorylation of serine residue 10 and acetylation of residue 14 on H3 is a tell-tale sign of transcription.

Chromosome remodeling through histone variants: 'H3 barcode hypothesis': In addition to post-transcriptional modifications (PTM) of histones, a number of variants are known for histone proteins that are found in the core particle of the nucleosome (Henikoff et al., 2004).

Perhaps H4 is the only exception, which does not seem to have any variant, but H3 and H2A are the two major classes of histones, which exhibit higher variation relative to H2B histone. Each histone variants differs from its corresponding normal histone in only a few amino acid residues (for a review, see Pusarla and Bhargava, 2005), and each histone variant plays an important role in chromatin remodeling.

In particular, the histone variants for H3 offer the best example to illustrate the role of histone variants in chromatin remodeling, so that a H3 barcode hypothesis has also been formulated recently to describe the role of histone H3 variants (Hake and Allis, 2006).

The histone variants are generally synthesized directly from genes encoding them, but may also result due to post-transcriptional modification of conventional histones.

The genes encoding histone variants can be broadly classified into replication dependent (RI), replication independent (RI) and tissue-specific (TS), so that at least some of these variants occur in a tissue specific or developmental stage-specific manner. These histone variants replace the normal histones, and can be incorporated into the chromatin at any time during the cell cycle, although the conventional nucleosomes are produced and assembled into nucleosomes only during the S-phase of the cell cycle.

WHOLE GENOME SEQUENCING : GENOMICS AND EPIGENOMICS

In the year 2000, with the publication of the whole genome sequence of the crucifer weed, thale cress (Arabidopsis thaliana), plant cytogenetics entered into a new era of research, the era of plant genomics. Consequently in the early years of the present century, whole genome sequences of rice and a draft sequence of the genome of poplar (Populus trichocarpa) became available. Sequencing of several other plant genomes is also in progress (as listed at NCBI site)

According to some, the first wave of plant genome sequencing is over, and we are now entering a new era in plant genomics research. In this new era, genomes of many model species with small genomes or those of species of economic importance will be sequenced.

Also the available sequences will be subjected to annotation (assigning functions to these sequences), and the choice of new genomes to be sequenced will be made on several criteria, including phylogeny (Jackson et al., 2006). The genomes of crops like maize and wheat will also be subjected to identification of gene-rich regions (GRRs), which will then be taken up for sequencing. There are also new technologies that will change the way we approach future genome sequencing projects.

As discussed earlier in this chapter, DNA methylation, nucleosome remodeling (including histone modification and histone variants), and noncoding RNAs can organize chromatin into accessible ('euchromatic') and inaccessible ('heterochromatic') sub-domains. This extends the information potential of the genetic code, and one genome can generate many 'epigenomes' in time and space, during the life-span of an organism.

The implications of epigenetic research are far reaching, so that efforts are being made to study the epigenomes in a variety of eukaryotes including some plant systems. In a recent study, it was shown that these epigenetic modifications are not as conserved as was once thought. Further, very little is known about histone methylation in large genome plants (Houben et al., 2003), which make up the bulk of the angiosperms (Arumuganthan and Earle, 1991).

4

Transgenics: An Emerging Approach

By the 2020, global population is expected to reach 8 billion. Conventional breeding methods alone cannot feed the extra hungry mouths. Despite the successes of the green revolution with substantial strides in food grains production, India is still classified by FAO as a low income, food deficit country, nearly 26per cent of India's population (268 million) are considered food insecure, consuming less than 80per cent of minimum energy requirements and account for about a third of the world's population that suffers from chronic hunger. Agricultural biotechnology has the potential to reduce crop losses from pest and diseases, to improve the nutrient efficiency of food and animal feeds; to extend post harvest losses with increased shelf life of fruits and vegetables; and to increase the stress tolerance of crop plants allowing them to tolerate various environmental extremes such as cold and drought. The biotechnology has the potential to enhance the productivity and also extend the cropping period especially in high altitude areas where vegetable cultivation is impossible in severe winter months. Modern Biotechnology has the tools to develop cold tolerant varieties, which enhance the productivity and profitability of farming. It empowers the hill farmers by boosting the productivity and also enhancing their nutritional security.

The present chapter examine the use of cold tolerant genes in vegetables and also tools of biotechnology the DRDO agricultural laboratories employs towards transformation of hill agriculture through development of cold tolerant transgenics in vegetables.

INTRODUCTION

Freezing temperature greatly limits the geographical distribution of cultivated plants and often causes severe losses in agriculture production and productivity. Conventional breeding methods have met with limited success in improving the freezing tolerance of important vegetables. Biotechnology offers new strategies that can be used to develop transgenics crop plants with improved tolerance to freezing stress. A number of genes have been isolated and characterized that are responsive to freezing stress. The molecular tool makes it possible to select directly at the gene label without waiting for the phenotype to show up. Therefore it is important to use most appropriate tools that help in reaching the goals.

The designed genotype should be better than the available ones and must reach the farmers. Transgenic technologies have opened up many exciting possibilities to improve products with added value with application in food, agriculture, animal husbandry, environment, medicine and industry. It also offers uncommon opportunities for improvement in genetic potential of plants and animals by introduction or removal of gene or genes that regulate a specific trait. The conventional breeding methods can be complimented by an array of bio-technological tools can be used to augment vegetable production by saving time and resources. The potential outcome be in the form of development of specific vegetable varieties that are more resistant to biotic and abiotic stresses, enhanced nutritional level of food items, enhanced shelf life of perishable farm produce, conversion of organic waste into bio-fuels.

As 100 million new people are added to the world's population each year and world population is predicted to reach eight billion by the year 2010. The population increase in developing countries constitutes 97 percent of the global increase. Feeding 3 billion additional people will require dramatic increase in crop production, a formidable task by any standard. There are limitations to conventional plant breeding technology either due to the limited gene pool or the restricted range of plant species between which genes can be transferred due to species barriers. Resource-poor farmers carry out 60 percent of global agriculture, but produce only 15-20 per cent of the world's food. The crops face major challenges from abiotic and biotic stresses and limited access to external inputs like pesticides, fertilizers and irrigation. The newly acquired ability to transfer genes between organisms without sexual crossing provides breeders with new opportunities to improve the efficiency of production and increase the utility of agricultural crops. While the new tools, technologies, and products will come from rapid advances in molecular biology and genetic engineering. The rapid development of plant biotechnology, agriculture can be efficiently moved from a resource based to a science-based industry.

PRESENT STATUS OF TRANSGENIC CROPS

The global area planted with transgenics crops has risen from 1.7 million hectares in 1996 to 90 million hectares in 2005 (James, 2005). Out of the total area of the world under transgenic crops, three countries namely USA, Argentina and Canada, which cover more than 90 percent.

The major crops occupying about more than 80 per cent area are herbicide-tolerant soyabean , Bt corn, Herbicide tolerant canola, and herbicide tolerant cotton . The three varieties namely Mech 12, Mech 162 and Mech 184 of Bt cotton have been released in India for commercial cultivation. Several transgenic cultivars of major food crops, such as soyabean, maize, canola, potato and papaya have been commercially released incorporating genes for resistance to herbicides, insects, and viruses.

Transgenic crops under commercial production across the globe.

According to (Manjunath, 2005), 04 field crops, 03 vegetables, 01 fruit crop and 01 tobacco crops are presently produced commercially across the globe is depicted in table 1.

Type of crop	Name of crop	Name of country
1. Vegetables	Tomato & Pepper	China
	Squash	USA
2. Field crops	Soyabean	Canada, USA, Argentina, South Africa, Brazil, East Europe, Uruguay, Paraguay and Chile
	Cotton	USA, Australia, Argentina, Mexico, China, South Africa, India and Colombia
	Maize	Canada, USA, Western Europe, Argentina,South Africa, Uruguay, Phillippines and Chile
	Canola	Canada and USA
3. Fruits	Papaya	Hawaii (USA)
4. Other crops	Tobacco	USA

Till date no transgenic crop variety tolerant to abiotic stress has so far been reported to be released for cultivation.

However, 7 transgenic varieties exhibiting tolerance to a range of abiotic stresses underwent field testing in Bolvia (frost tolerant potato variety), China (cold tolerant tomato), Egypt (salt tolerant wheat variety), India (Moisture stress tolerant Brassica variety) and Thailand (salt tolerant and drought tolerant rice varieties.

Most of the R&D activities on development of abiotic stress tolerant crops are being carried out in six countries of the Asian region namely Bangladesh, China, India, Indonesia, Pakistan and Thailand.

MAJOR COMPONENTS FOR THE DEVELOPMENT OF TRANSGENIC PLANTS

Despite significant advances over the past decade, development of efficient transformation methods can take many years. Effective regeneration and transformation systems are the prerequisite for successful genetic transformation (Sharma et al, 2005). Genetic transformation technology relies on the conceptual framework and the technical approaches of plant tissue culture and molecular biology to develop commercial process and products which comprises mainly the nine major components:

(i) The development of reliable tissue culture regeneration systems

(ii) Preparation of gene construct and transformation with suitable vectors

(iii) Efficient techniques of transformation for the introduction of genes into crop plants

(iv) Recovery and multiplication of transgenic plants.

(v) Molecular and genetic characterization transgenic plants for stable and efficient gene expression

(vi) Transfer of genes to elite cultivars by conventional breeding methods if required

(vii) Evaluation of transgenic plants for their effectiveness in alleviating the biotic or abiotic stresses and field performance

(viii) Bio-safety assessment including food and environmental safety

(ix) Commercialization of genetically engineered crop.

IMPORTANT CROPS DEVELOPED THROUGH GENETIC ENGINEERING

A number of transgenic plants developed so far through plant genetic engineering for different agronomic and qualitative traits are:

(i) Herbicide tolerant transgenic plants have been engineered by using mutants EPSP (5-enolpyruvyl shikimate-3-

phosphate) synthase enzyme e.g. soybean. Canola, cotton and tobacco.

(ii) Transgenic for insect resistance have been developed introducing a gene from Bacillus thuringiensis coding a toxic protein (delta endotoxin) which inhibits insect growth. e.g. Bt cotton, Bt maize, Bt tobacco.

(iii) Transgenics for male sterility (Barnase and Barstar in Brassica spp. for hybrid seed production).

(iv) Stress resistance transgenics. e.g. chilling resistance in tobacco, a gene for glycerol-1- phosphate transferase from Arabidopsis.

(v) Transgenics for disease resistance such as fungal diseases, bacterial diseases, viral diseases.

(vi) Transgenics for food processing, Bruise resistance tomato (expresses antisense RNA against

polygalacturonase which attacks pectin), delayed ripening

HILL AGRICULTURE AND PROSPECTS OF COLD TOLERANT GENES

Cold is an environmental factor that limits the geographical distribution and growing season of many plant species, and it often adversely affects crop quality and productivity (Thomashow, 1999). Most temperate plants can acquire tolerance to freezing temperature by a prior exposure to low nonfreezing temperature, a process known as cold acclimatization (Guy 1990, Hughes and Dunn,1996, Browse and Xin, 2001).

Plants of tropical and subtropical origins are sensitive to chilling temperature (0°C-100°C) and are incapable of cold acclimation. Many studies have suggested that cold regulated gene expression is critical in plants for both chilling tolerance a (Gong et al. and Hsieh et al. 2002) and cold acclimation (Knight et al. 1999: Thomashow 1999, Tamminen, 2001).

Cold responsive genes encode a diverse array of proteins such as enzymes involved in respiration and metabolism of carbohydrates, lipids, phenylpropanoids and antioxidants: molecular chaperones, antifreeze proteins, and others with a presumed function in tolerance to dehydration caused by freezing (Guy 1990, Thomashow 1999 and Mohapatra et al. 1989).

The change in the gene expression occur in plant during cold acclimatization a developmental process that results in increased tolerance (Guy, 1990 and Steponkus et al. 1993). Since then, it has repeatedly been speculated that certain COR (cold regulated) genes might have role in freezing tolerance. To test this notion investigators have turned to isolating the characterizing genes that are expressed in response to low temperature.

These efforts have led to the identification of a number of genes such as the COR 15a KINI LTI 78 fad 7 genes of Arabidopsis thaliana. It is well known to farmers and scientists that low temperatures can kill plants. Low temperature stress is a major environmental factor that not only limits where crops can be grown but also reduces yields depending on the weather in a particular growing season.

In addition to exceptionally stressful years that cause significant yield reductions, less extreme stress almost certainly causes smaller losses over large areas to produce comparable yield reductions every year. Even in cases when freezing stress does not result in yield loses., it often results in crop quality reduction.

Each year, worldwide losses in crop production due to low temperature damage amount to approximately $ 2 billion. Some of the major losses include the 1995 early fall frosts in the US which caused losses of over $1 billion to corn and soyabeans. The occasional freezes in Florida have shifted the citrus belt further south, and California sustained $650M of damage in 1998 to the citrus crop due to a winter freeze.

The inherent cold hardiness of the crop determines in which agricultural areas it can be grown. Crops that are more resistant to freezing stress would allow some geographical regions to grow more profitable and productive crops with less environmental risks. However despite continued efforts, traditional breeding has had only limited success in imparting crop plants with better freezing tolerance due to very little was known about the mechanisms that regulate chilling and freezing tolerance. With the advent of molecular genetics and biotechnology, it is now possible to genetically engineer plants to be more tolerant to many environmental adversities, including low temperature.

COLD ACCLIMATIZATION AND SIGNAL TRANSDUCTION CASCADE

Plants vary greatly in their ability to with stand low temperature stress. As an adaptation strategy, most plants native to the temperature climates develop freezing tolerance after prior expose to cold, Non-freezing temperatures, a phenomenon called acclimatization. Many biochemical and physiological changes are known to occur during cold acclimation. With the onset of low temperature, putative temperature sensors at the cellular membrane generates stress signals which are transmitted and amplified through many steps that include Ca++ signaling and a stepwise kinase/phosphatase chain reaction termed the kinase cascade.

The massage eventually reaches the nucleus and regulators of gene expression called transcription factors, which act as master switches to regulate the expression of groups of genes, resulting in the increase of proteins and other organic molecules that protect the cell from freezing damage. The plants response to cold acclimation is clearly complex and diverse, and the actual biochemical and physiological changes are still poorly understood at the molecular level.

Technology Development toward cold Tolerance

The cbf genes represent one of the most significant discoveries in the field of low temperature adaptation and signal transduction. All-important crops and few vegetables and tree species have contained this gene. The transgenic canola contain this gene are able to survive freezing temperature as much as 4-50C lower than the non transgenic controls (Jaglo et al. 2001). Tomato plants have also been successfully engineered using the CBF genes to achieve chilling tolerance (Hsiech et al. 2002).

Thus the CBF technology has a great potential for improving the cold and freezing tolerance in plants. There is strong evidence that the decrease in fatty acids saturation that generally occurs in temperate plants upon exposure to low temperature contributes the cold tolerance of these plants.

Ishizaki-Nishizawa et al. 1996 found that by introducing a broad-spectrum desaturase gene from a cyanobacterium into tobacco plants could increase the low temperature tolerance of the transgenic plants. Expression of a plant phosphatase (At PP2CA) in transgenic Arabidopsis thaliana can accelerate the development of cold acclimation and increase freezing tolerance.

It has also shown that transgenic plants expressing a constitutively active kinase NPK1, is more tolerant to chilling and other abiotic stresses (Kovtun et al. 2000). Other transcription factors, including ABI3 and SCOF-1, have been used to successfully increases the cold tolerance of transgenic plants (Tamminen et al. 2001). Transcription factor genes Abl3, abi3, cbf1, dreb1A, dreb1 and dreb2 from A. thaliana, osmolyte biosynthesis genes from Arthrobacter globiformis and afb (anti freeze protein) from A. thaliana are some other important genes which have effective role to cold tolerance in plants.

GENES USED FOR TRANSGENIC DEVELOPMENT IN TOMATO, CAPSICUM AND CUCUMBER FOR HIGH ALTITUDE REGION

Advancement in plant biotechnology has led to the identification and isolation of a number /transcription factor(s) related to abiotic stress tolerance. Defence Research and Development Organisation has taken up the work on development of transgenics for cold tolerance through cloning and transformation of these genes into high value vegetable crops to cultivate them in higher altitude areas for the fresh requirement of armed forces ane local populace. At present the work is carried out to develop transgenics with the following genes in tomato, cucumber, pea and brinjal and cloning of cold tolerant genes from seabuckthorn plant.

Osmotin gene: Among the various environmental factors, availability is the major factor influencing growth, development and productivity of crop plants.

Accumulation of low molecular weight osmolytes such as praline, betaines and sugar alcohol is an important mechanism underlying adaptation to such stress factors. In addition, stress induced proteins whose functions, however, remain largely unknown and enzymes that toxify reactive oxygen species can also contribute to stress tolerance. Osmotin is a basic 24 KD protein that was originally identified in tobacco adapted to NaCl and desiccation.

The stress induced synthesis and accumulation of the osmotin protein is correlated with osmotic adjustment in tobacco cells. The synthesis of osmotin is induced by ABA that accumulates in response to osmotic stress and subsequently plays a pivotal role in osmotic adjustment.

It is known that expression of osmotin gene induces proline accumulation in unstressed and stressed plants and imparts tolerance to both salinity and drought stress. The gene has been transferred through Agrobacterium in tomato at DARL, Pithoragarh and transgenic plants are obtained and T3 plants are being tested in high altitude areas under controlled environments.

Mannitol-1 phosphate Dehydrogenase (mtlD): Mannitol-1 phosphate dehydrogenase (mtlD) is one of the gene encoding enzyme that synthesize osmoprotectants and enhance their expression in transgenic plants which leads to maintenance of osmotic potential during the stress period . The candidate gene has been isolated from

E. coli K-12 strain and code for an enzyme called mannitrol-1 phosphate.Mannitol-1 phosphate then converted into mannitol by non specific phosphates. Accumulation of mannitol in different parts of a plant leads to tolerance at cellular level by adjustment of the cytosolic osmotic potential when the concentration of electrolytes is lower in the cytosol than in the vacuole.

OSISAP1 gene of rice and CBF1 gene of Arabidopsis: The expression of these genes changes in the membrane lipid composition, accumulation of compatible osmolytes or cessation of plant growth. Cloning and expression of these genes has led to the development of strong tolerance to freezing through transgenic approaches. Dehydration responsive element binding factor (DREB/CRB3): The gene is responsible for the expression of many cold tolerant genes during cold stress in plants.

Glyceraldehydes phosphate Acetyl Transferase (GPAT) gene: The gene is responsible for the un saturation of the fatty acids present in the plant cell wall, which give the cold tolerance to the plants during cold stress. Cold regulated (COR15A) gene: This gene encodes for polypeptides that decreases the incidence of freeze induced lamellar to hexagonal II phase transitions, which increases freezing tolerance to the plants. Desaturase (desC) gene: The expression of this gene increases the fatty acid unsaturation of the cell wall lipids giving strength to the cell wall of the plant during cold stress.

CONCLUSION

The high altitude regions have a very harsh climate and a short agriculture season. Only a few vegetable crops viz, radish, turnip, potato, cabbage and some leafy vegetables are grown traditionally, in these areas. The climatic and geographic differentiation segregates the region from rest of the world.

Therefore, the agro-techniques for the vegetable production being employed elsewhere is not suitable for these conditions and specific agro-techniques and vegetable varieties are required for these regions. Lots of works has already been done to identify the suitable varieties/ hybrids and cultivation practices for these regions. The transgenic technology is highly precise and beneficial for high altitude farming. The major limitation is the complexity of tolerance to abiotic stresses that is normally dependent on a number of physiological traits, each under multigenic control.

The increased wealth of knowledge that is being acquired by means of genomic and other molecular biology studies and the cloned genes will certainly contribute to the development of tolerant genotypes. The successful integration of the cold tolerance genes into high value vegetable crops would increase productivity and production in high altitude regions. The transgenics having cold resistance gene may not only bring additional areas under cultivation but also help in optimizing productivity in high altitude and remote areas without any additional cost will enable or extend the cultivation of crops in abiotic stress, which are presently not under cultivation or are less productive.

5

MOLECULAR CYTOGENETICS IN PLANT IMPROVEMENT

Researchers need to utilize new molecular genetic information about cereal species and the accompanying technologies to produce new information and genetic materials. This project will increase the efficiency of genetic analysis, suggest tests that relate specific genetic regions to productivity, develop new genetic lines that cause a shift in microbial populations in animal digestive systems, and improve the nutritional value of crops.

OBJECTIVES

Major genome research programs are underway to isolate expressed genes of corn and to DNA sequence at least the gene-rich regions. We are developing a highly efficient method for mapping DNA sequences to chromosome and sub-chromosomal regions based on lines derived from crosses between oats and corn. Partial hybrids have been obtained that contain an entire set of oat chromosomes along with one corn chromosome; lines are available that have each chromosome of corn individually added to the oat genome.

These lines are now being used to generate derivatives with only a subregion of the corn chromosome. This genetic material as well as B73xMo17 derived lines will be used to learn more about the subgenomic structure of corn and to test whether the degree of heteroallelism within and between the subgenomes contribute to heterosis. A pathogenic bacteria, called E. coli O157:H7, can contaminate hamburger and cause serious hemoraging. The contamination largely comes from the feces of cattle. Strains of E. coli have been selected that will kill the O157:H7 strain.

Two genes appear to be involved and this project will test their efficacy in corn. Wild rice is a crop under domestication. The seed falls (shatters) from the panicle upon maturation and is not available to be harvested. Storms can cause major losses of wild rice yield at harvest.

The shattering trait is considered a domestication trait that has been modified years ago in most major crops; the trait still exists in wild rice varieties. We have discovered major genes for shattering and have molecular genetic markers that should allow more efficient breeding for shattering resistant varieties. Methionine is an essential amino acid required in the diets of non-ruminant animals. Poultry have high demands for methionine accounting for why synthetic methionine is added to poultry rations in the U.S. A line of corn was selected that is high in methionine. Several generations of crossing, backcrossing, and selfing has resulted in relatively homozygous lines potentially high in methionine,

The specific objectives are:

1. Develop efficient mapping system for corn DNA sequences.
2. Learn more about the subgenomic structure of corn and relate the information to productivity.
3. Manipulate corn to produce compounds toxic to an enterohemoraghic bacterium (E. coli O157:H7).
4. Provide molecular genetic methods to the breeding program to enhance the selection of varieties of wild rice with improved seed shattering resistance.
5. Finalize the selection and release of high methionine lines of corn.

We have produced lines of oats with a single corn chromosome added. These addition lines are available for all 10 corn chromosomes, individually present in the oat background. These lines are being used extensively for mapping DNA sequences to chromosome.

The oat-corn addition lines are being irradiated to produce radiation hybrid (RH) lines that have only part of the corn chromosome present. These lines allow much more refined mapping of corn DNA sequences. Future research will involve continued development and characterization of the RHs, mapping ESTs, cytogenetic transmission studies of the various lines, and obtaining the transmission of the chromosome 10 addition and formation of RHs.

Molecular genetic analysis of corn has shown that its genome is composed of major duplicated regions suggesting an ancient allopolyploid origin.

The potential exists for many loci to not only be duplicated but to have different alleles at each locus. Maximum productivity could be the result of maximizing the number of alleles at these loci. We will screen several genetic markers in order to determine Mo17xB73 lines containing duplicate regions containing various combinations of the parental genotypes.

Yield tests will ultimately be performed on selected lines with maximum heteroallelism in regions known to possess major yield QTLs. Our molecular genetics research on wild rice has resulted in the first molecular marker map and the placement of QTLs for several traits on the map. A PCR based marker has been developed that tags the most major QTL with polymorphism between shattering and non-shattering types.

This QTL accounts for about 40% of the variation in seed shattering in wild rice. A marker-assisted selection program will be initiated using the molecular genetic marker for the major QTL and others as possible. Escherichia coli O157:H7 is a pathogenic strain of E. coli found in the digestive tract of beef cattle and causes food borne illness through contamination of beef during the slaughtering process. The goal of this project is to target the pathogen at the feed source or farm level to prevent contamination by developing a transgenic maize line that expresses a protein (termed "colicin") toxic to this strain of E. coli.

Colicin (produced by the cea gene) degrades DNA, and an immunity protein (produced by the cei gene) is required to prevent non-specific DNA degradation. Thus both genes are required to accomplish our goal. A screening procedure was developed in our lab that provides a screen for corn with elevated levels of methionine in the endosperm. A line was discovered, called BSSS5, that is resistant to lysine-threonine inhibition as a whole kernel but is inhibited when only the embryo is cultured.

The endosperm in this case possesses sufficient methionine to prevent the inhibition. BSSS 53 was found to produce a zein-2 protein with 23% of the amino acid residues as methionine. We have attempted to backcross the high-methionine trait into three different genetic backgrounds (B73, Mo17, and A632) and have some lines almost ready to release.

Oat-corn addition lines have been developed for every corn chromosome. These strains allow - in a single experiment - the rapid mapping of corn DNA sequences to their respective chromosome.

These strains resulted from more than 100,000 oat x corn crosses and isolation of over 7300 immature embryos. We extended the available genetic backgrounds by recovering oat-corn addition lines with B73 or Mo17. Between the two inbreds, we have a complete addition line set; the B73 addition lines will be useful in assembling the corn genome.

The DOE Joint Genome Institute will sequence chromosome10 using our Mo17 chromosome 10 oat-maize addition line. We also have added to our collection of radiation hybrid lines for each chromosome. These lines allow the placement of corn genes to specific regions of the chromosome. Corn gene expression/silencing in the oat-corn addition lines is being tested using the DNA chip technology. Out of 17,000 tested sequences, 890 were expressed in the chromosome 5 addition.

About half of these sequences are of known position and are either placed on chromosome 5 or in a duplicated region of the genome. We further pursued the possibility of introducing C4 photosynthesis to oat with these materials.

Addition lines with both chromosomes 6 and 9 (carrying genes for key C4 enzymes) were tested for an effect on CO2 compensation point. A statistically significant change occurred in compensation point but not an impressive one. We will next investigate a third chromosome that has an effect on leaf cell morphology (known as Kranz Anatomy) involved in photosynthetic efficiency.

NIR analyses of self-pollinations of a new high oil accession indicate that the kernels contain 20% oil. A QTL analysis is underway. There is considerable interest in this material as a source of biodiesel and ethanol.

Using the molecular genetic map we published for wild rice, a PCR-based marker system has been identified for use in marker assisted breeding for the seed shattering trait. Genetic tests indicate a close marker/trait linkage.

The recent outbreak of the pathogenic E. coli 0157:H7 on spinach emphasizes the importance of having control measures to minimize such occurrences. We are developing what might be considered a prototype approach of using corn feed to alter the microflora in the intestinal tract as a means to reduce the presence of this pathogenic bacterium at its source.

Whole corn plants were regenerated from appropriate calli and found to carry the colicin gene, express the RNA and protein, and have anti-microbial activity against E. coli O157:H7. An extensive bioinformatics analysis has been performed of 1100 DNA sequences involved in human cancers to determine if they are present in plants. P53 is a protein involved in more than 50% of human cancers. We found a p53 binding sequence in the promoter of the MAD1 (Mitosis Arrest Deficiency) gene in human; MAD1 acts as a mitotic checkpoint to ensure chromosome stability. A MAD1 homolog was found in maize. Laboratory analyses of knockout mutants of MAD1 in Arabidopsis are underway.

Crop productivity is at least half due to genetic advances. Continued understanding of genetics and the application of molecular genetics will allow continued progress in food production. In addition, additional health benefits and identification of unique biofuel resources will be forthcoming.

6

Identification of Polytene Chromosomes

INTRODUCTION

Since the discovery of polytene chromosomes in the embryo suspensor of *Phaseolus coccineus* (Nagl 1962a) attempts were made to accurately describe each chromosome (Nagl 1962b, 1965). A preliminary idiogram based on total chromosome length and the distribution of euchromatic and heterochromatic segments was first presented by Nagl (1967). Later, Nagl (1974) pointed out, however, that the length can vary considerably depending on the level of endopolyploidy, the state of condensation and the degree of stretching during preparation. Therefore it was necessary to improve the description of the karyotype and to find a more reliable and constant feature in order to identify each chromosome of a given nucleus. We report here the first results of the identification of the polytene chromosomes of *P. coccineus* based on the analysis of centromeric heterochromatin morphology after pepsin pretreatment and DAPI staining.

MATERIAL AND METHODS

Seeds of *Phaseolus coccineus* cv. Preisgewinner were grown in the Botanical Garden of the University of Kaiserslautern. Embryo suspensors were taken from seeds of unripe pods and preparations of polytene chromosomes were made from the biggest basal suspensor cells (Nagl 1996). After enzymatic maceration the nuclei were gently squeezed out from single cells to get rid of the cell wall. Additionally, the extracted nuclei were incubated in 45% acetic acid before squashing in order to dissolve the cytoplasm to some extent, and thus improving the spreading of chromosomes. After removing the cover slip and air drying, the preparations were pretreated in the same way as was carried out for fluorescence *in situ* hybridization (Nenno et al. 1994). Chromosomes were then stained with DAPI. Pictures from chromosomes of ten well preserved nuclei were taken by a cooled CCD camera at high resolution and analyzed by using an image analysis system (IPLab Spectrum).

RESULTS AND DISCUSSION

The analysis of polytene chromosomes from the embryo suspensor of *P. coccineus* after pepsin pretreatment and DAPI banding made it possible to give a better description of individual chromosomes. The new key feature is the centromeric heterochromatin morphology (CHM). Two characteristics make the CHM useful for chromosome identification. Firstly, it is less sensible to variations in level of endopolyploidy and the degree of stretching during preparation. Secondly, the CHM of each of the 11 chromosome pairs can be individually identified by its pattern of at least one bright DAPI band as well as by the position of the centromere and the presence of secondary constrictions.

For the construction of an idiogram, traditionally, the total length and the ratio of the short and the long arms are relevant. In the polytene chromosomes of *P. coccineus* these parameters, however, vary significantly due to the state of condensation and the degree of stretching during preparation. Especially, the euchromatin behaves like an elastic band. For this reason, the arrangement of chromosomes labeled C-J is arbitrary, and the euchromatic parts are marked with dashed lines. Theses chromosomes are metacentric to submetacentric. Chromosome A is the longest and is telocentric. B is the next longest chromosome and the only one which is generally submetacentric. Chromosome K is the smallest of the complement and is acrocentric.

Besides the new characterization of the chromosomes by the CHM there are two striking differences between the new idiogram and that previously proposed (Nagl 1967). The new idiogram shows three chromosome carrying a NOR. This was also demonstrated by silver-staining (Schweizer and Ambros 1979) and by fluorescence *in situ* hybridization (Nenno et al. 1994). Furthermore, the existence of the formerly named 'SAT-chromosomes' 1 (S_1) and 5 (S_2) (Nagl 1965, 1967) could not be confirmed. Our present results indicate, that chromosome fragments sticking to the nucleolar material were obviously misinterpreted as 'chromosome satellites'. These fragments seem to arise from the small centromeric heterochromatin block close to the NOR on the short arm of chromosome K, from which the long arm readily detaches.

The problem, however, that not all of the eleven pairs can be recognized equally well is not yet solved. Although, eight pairs of homologous can now be recognized quite easily, the identification of chromosome H, I, and J is more difficult. Nevertheless, the preliminary results presented here are encouraging and suggest that in a near future it will be possible to produce a new, high resolution karytotype for *Phaseolus coccineus*.

Hatched boxes:NORs; *Triple dots*: Telomeric heterochromatin: *Horizontal hatching*: Centromeres which are not always visible; *Dark* : Euchromatin; Centromeric *Dark gray*: heterochromatin (CH); *Gray* or *white*: DAPI-bands within the CH, according to their relative staining intensity. *Square brackets*: structures which are not always present; *Double headed arrow*: the position in K, where the long arm detaches readily; *Asterisks*: regions which normally are wider than the rest of the chromosome. Bar represents 10 µm.

CYTOGENETIC STUDIES IN *PHASEOLUS* L. (FABACEAE)

A review of the cytogenetic studies carried out on *Phaseolus* as well as the different proposals that have been suggested to explain the chromosomal changes in the group are presented. The importance of including wild species in cytogenetic studies and the collaboration between taxonomists and cytogeneticists in order to draw better conclusions are emphasized.

INTRODUCTION

The family *Fabaceae* (*Leguminosae*) contains the sub-family *Papilionoideae* of which the tribe *Phaseoleae* is one of the most important groups because it contains genera such as *Glycine* (soybean), *Phaseolus* (American beans) and *Vigna* (Asiatic beans), which are economically important due to their role in human nutrition and their use as cattle forage and ornamental (Lackey, 1981).

The genus *Phaseolus* is mainly found in the Mexican mountains (Sousa and Delgado, 1993), and contains approximately 50 species, with four (Delgado-Salinas, 1985) or five (Debouck, 1991) cultivated ones: *P. vulgaris*, *P. coccineus*, *P. acutifolius*, *P. lunatus* and *P. polyanthus* (= *P. coccineus* subsp. *darwinianus*).

In his original description of *Phaseolus*, Linnaeus (1753) included eleven species, but with time the number grew to 200, distributed both in the Old and the New World. In 1970, Verdcourt redefined *Phaseolus*, considering it exclusively of New World origins, with approximately 50 species whose characteristics are similar to those of *P. vulgaris* (generitype). This redefinition was confirmed as valid and refined by a series of studies by other researchers (Maréchal *et al.*, 1978 and Lackey, 1981, 1983).

The last revision of the genus was made by Delgado-Salinas (1985), who recognized only 36 species in North and Central America. Despite the taxonomic studies carried out on the genus that have led to its clear delimitation, neither the number of taxa of which the genus is composed nor the genetic relationship between species has been well established (Debouck, 1991). Delgado-Salinas (1985) estimates that the genus contains 36 species, in North and Central America, some of them with subspecific divisions, while Debouck (1991) includes 52 especies, without subspecific divisions. Although the importance of cytogenetic studies have been noted by several authors (Thomas, 1973; Green *et al.*, 1980; Almeda and Chuang, 1992), most studies have dealt with economically important species, ignoring the potential of wild species and relating only to cultivated species such as *Phaseolus*.

CHROMOSOMAL STUDIES

The first reports on chromosome numbers in *Phaseolus* go back to 1925, when Karpetschenko obtained $2n = 22$ for *P. acutifolius* A. Gray, *P. coccineus* L., *P. lunatus* L. and *P. vulgaris* L. From then on, a large number of cytogenetic studies have focused mainly on the determination of chromosome numbers, establishing $x = 11$ as the basic number.

Prior to 1996, of the approximately 50 species recognized in the genus *Phaseolus*, only 9 species and 4 subspecies had been chromosomally counted. Mercado-Ruaro and Delgado Salinas (1996, 1998) increased the number of taxa analyzed to 31. Based on the published literature Lackey (1979), Goldblatt (1981) and Mercado-Ruaro and Delgado Salinas (1996, 1998) propose that, as in the tribe *Phaseoleae*, the basic chromosome number in the genus is $x = 11$, with a haploid number of $n = 10$ in three species (*P. leptostachyus* Benth., *P. micranthus* Hook. & Arn., and *P. macvaughii* A. Delgado, ined. (Mercado-Ruaro and Delgado-Salinas, 1998)).

The number of species that have been analyzed is very low, and the analyses have been restricted mainly to cultivated species (Sarbhoy, 1977; Joseph and Bouwkamp, 1978; Sinha and Roy, 1979a; Zheng *et al.*, 1991). Mercado-Ruaro and Delgado-Salinas (1998) reported the karyotypic analysis of 10 wild species, that represent on average 20% of those comprising the genus. The lack of karyologic studies in the genus has been attributed to the reduced size of the chromosomes, which makes the analysis difficult (Hucl and Scoles, 1985; Zheng *et al.*, 1991). Nonetheless, the available information has shown that there is a predominance of metacentric and submetacentric chromosomes, which translates into very symmetrical karyotypes.

Some authors (Sarbhoy, 1977, 1980; Sinha and Roy, 1979a,b) have pointed out that the main factors involved in the karyotypic evolution of the genus are pericentric and paracentric inversions, translocations and the loss or gain of chromatin. They have also proposed that the karyotype of *Phaseolus* has evolved towards an asymmetry, with a decrease in the total chromatin content. Mercado-Ruaro and Delgado-Salinas (1998), after encountering three aneuploid species with $2n = 20$, have pointed out that aneuploidy has also played a role in the evolution of the karyotype.

GENOMIC HYBRIDIZATION

Studies of *Phaseolus vulgaris* by Frediani *et al.* (1993) using *in situ* hybridization have shown the position of the genes that code for polygalacturonase-inhibiting protein (PGIP) and stablished that the coding sequences are located in the heterochromatic pericentromeric region of metacentric chromosome 10, while Schumann *et al.* (1990) and Nenno *et al.* (1993) have documented the position of the phaseolin gene. In *P. coccineus* Avanzi *et al.* (1972) have located the ribosomal cistrons in the nucleolar and satellite regions of chromosomal pairs I and V using tritium-labelled rRNA. These studies, all employing polytene chromosomes, show the potential of *in situ* hybridization for chromosome mapping.

The application of genomic *in situ* hybridization to taxonomy and the elucidation of genetic relationships are exemplified by the studies of Mercado-Ruaro on the *Phaseolus vulgaris-P. coccineus* complex, which investigated the possible hybrid origin of *P. coccineus* subsp. *darwinianus* Hernández X. & Miranda C. (= *P. polyanthus* Greenm.) as well as the genetic relationships between the species and subspecies that make up the complex. The results of this study have shown the high degree of genetic homology between the members of this group, and because of this it was not possible to establish whether or not *P. coccineus* subsp. *darwinianus* is the result of a cross between *P. coccineus* and *P. vulgaris*, although it was possible to establish that *P. glabellus* is a taxon only distantly related to other members of the complex.

STUDIES OF NUCLEAR DNA CONTENT

There is much variation in the reported DNA content of the *Phaseolus* species studied by different authors. The DNA content of cultivated *P. vulgaris* has been reported as being 1.56, 1.63, 1.69, 1.79 pg by Castagnaro *et al.* (1990), 2.7 pg by Bennett (1982) and 3.7 pg by Ayonoadu (1974), while that of the wild-type *P. vulgaris* var. *aborigineus* has been reported to be 1.71 pg by Castagnaro *et al.* (1990). These differences may be attributable to the source of the material, the type of control used or to errors inherent in the technique. Other species studied for their DNA content include *P. coccineus*, containing 3.5 pg according to Ayonoadu (1974) and 1.98 pg according to Castagnaro *et al.* (1990); *P. lunatus* with 2.5 pg (Ayonoadu, 1974); *P. dumosus* with 3.8 pg, and *P. leucanthus* with 3.3 pg (Ayonoadu, 1974).

The latter two species are probably *P. coccineus* subsp. *darwinianus*, since both names have always been nomenclaturally associated with this subspecies.

As is the case in cytogenetic studies, there are reports of species referred as *Phaseolus* when they actually belong to other genera, for instance, *P. angularis* is really *Vigna angularis*, with a DNA content of 2.8 pg, while both species *P. geophilus* (2.6 pg) and *P. lathyroides* (2.3 pg) belong to genus *Macroptilium*.

Ayonoadu (1974) found a positive correlation between the nuclear DNA content and the nuclear volume, nucleolar and nuclear dry mass and total dry mass, i.e., high DNA content indicates high values for volume and dry mass parameters. Castagnaro *et al.* (1990), studying *P. coccineus* and several cultivars of *P. vulgaris*, along with *P. vulgaris* var. *aborigineus*, also found a positive correlation between seed weight and DNA content, with the exception of *P. vulgaris* var. *aborigineus*, which presented a negative correlation. Even so those authors conclude that varieties with a high DNA content are better adapted to cold or temperate regions, while those varieties with a lower DNA content are adapted to hot, dry environments.

We are now in the process of analyzing the DNA content of wild species of *Phaseolus* to determine if there is any relationship between DNA content and taxonomic relationships between the species and/or karyotype.

Plant Polytene Chromosomes

Polytene chromosomes are structures found in highly specialized tissues in some animal and plant species, which are amplified through successive cycles of endoreduplication, finally producing several copies of each chromosome. For this reason, they have been very important in elucidating chromosome fine structure and physiology, especially in diptera.

In plants, polytene chromosomes have been observed in only a few species, and seemed to be restricted to ovary and immature seed tissues, e.g., in *Phaseolus coccineus* and *P. vulgaris* (Nagl, 1981), until relatively recently, when they were observed in the cells of the anther tapetum of *Vigna unguiculata* (Guerra and Carvalheira, 1994) and of some *Phaseolus* species (Carvalheira and Guerra, 1994). With the discovery of the polytenics in tapetum tissue, it was observed that in many other species of various angiosperm families the tapetal cells also display polytene, polyploid or both types of nuclei. In some species of *Phaseolus* and *Vigna* the polytenics are more clearly defined and, therefore, better suited to the study of this type of chromatin organization. It is, however, important to differentiate between the nuclear cycles that result in polyploid nuclei and those that produce polytene nuclei, because these two terms of the nuclear types are often used indiscriminately in the literature. In this paper some aspects of the occurrence of plant polytenes will be summarized along with the structure and function of these chromosomes.

Endomitosis and Endoreduplication

Nagl *et al.* (1985) described polytene chromosomes as giant chromosomes produced by changes in the mitotic cycle during the interphase stage. In such a modified nuclear cycle, the chromatin duplicates its DNA content during the G_1 and S stages, but, instead of passing to the G_2 stage, the nucleus initiates a new G_1 phase, thus starting a new cycle of chromatin duplication. This type of cycle was first described in 1939 by Geitler, as occurring in the somatic cells of the insect *Gerris lateralis* (Painter and Reindorp, 1939; D'Amato, 1964), and was named the "endomitotic cycle" because it develops within the nuclear envelop without either achromatic spindle formation or nuclear or cellular division (Nagl, 1970a; Brodsky and Uryvaeva, 1985). The term endomitosis is, however, generally used to describe the formation of both polyploid and polytene nuclei (q.v. Nagl, 1974). Nagl (1978, 1981, 1987) has suggested the term endocycle rather than endomitosis, and D'Amato (1984) has adopted the term endomitotic and endoreduplication to distinguish between those that produce polyploid and polytene nuclei, respectively.

The endomitotic cycle (endomitosis) starts with a normal prophase (endoprophase), after which the chromosome contracts further (endometaphase), their sister chromatids separate from each other (endoanaphase) and decondense to assume the interphase nuclear structure, resulting in polyploid cells, with double the chromosome number (endopolyploidy) at the end of each cycle. The essential difference between endomitosis and the normal cell cycle is the absence of nuclear membrane dissolution in endomitosis, with the whole cycle occurring inside the nucleus. Such cycles have been observed in the anther tapetum of some angiosperm species, as in some *Passiflora* species and in *Papaver rhoeas*.

The endoreduplication cycle differs from endomitosis because it results in polytene cells (cells with many identical paired chromatids). In the endoreduplication cycle, the chromatid number is duplicated, but they do not segregate, and after various endoreduplication cycles, larger and thicker chromosomes are produced, called polytenics. In the endoreduplication cycle, the condensation and decondensation stages are not evident (D'Amato 1984, 1989), except in some cells where it is possible to see the chromocenter dispersion phase, known as the Z-phase (Nagl, 1970b, 1972; Cavallini *et al.*, 1981).

Depending on the behavior of the sister chromatids, polytene nuclei can be divided into two structural types. The first, and most well studied, are the chromosomes of the larval cells of *Drosophila*,

Chironomidae and other diptera (Ashburner, 1970; Brodsky and Uryvaeva, 1985). These polytenics are characterized by numerous transverse bands along their linear axis, produced by the exact pairing of sister chromatids and the intimate association of their chromomeres (Ashburner, 1970). The somatic pairing of homologous chromosomes gives the false impression that there has been a decrease in chromosome number, because each nucleus appears to contain the haploid number of giant chromosomes.

The other structural type of polytene nuclei also has the grouping of sister chromatid bundles resulting from several endoreduplication cycles, but in this case is characterized by the lack of any intimate pairing of the chromatids. This nucleus type is observed more frequently, typical examples being found in the giant trophoblast cells of mammals (Nagl, 1985), the trophocit cells of many insects (Painter and Reindorp, 1939), some ovary tissues during the development of many angiosperms (Corsi *et al.*, 1973; Nagl, 1976) and in the anther tapetum of some plant species (D'Amato, 1984; Guerra and Carvalheira, 1994; Carvalheira and Guerra, 1994, 1998). In these nuclei, which can be recognized both by the large size of their chromocenters and by the diploid number of polytene chromosomes, the chromosome number does not appear to be reduced as in polytene-type nuclei. Another peculiar giant chromosome type, which likewise does not present somatic pairing, has been found in some ciliates (e.g., *Stylonychia mytilus*) that have a macronucleus with polytene chromosomes and a diploid micronucleus (Ammermann, 1971; Ammermann *et al.*, 1974). The polytenics of these ciliates display band and interband patterns (also seen in *Drosophila*), but the macronucleus disintegrates after its development while the micronucleus remains active.

It is interesting to note that the endocycles are not processes of cell multiplication but are associated with cell differentiation and seem to be genetically controlled, with both endopolyploidy and polyteny leading to cell specialization in certain tissues. These nuclei have generally been observed in ephemeral tissues made up of only a few cells with intense metabolic activity, the main function of which is to provide nutritional support to vital organs during certain periods of development (e.g., the larval salivary glands of insects, the mammalian trophoblasts and the embryo suspensor cells of angiosperms). In such tissues, the cytoplasmatic volume and nuclei DNA content of the cells are increased by endomitosis or endoreduplication cycles (Nagl, 1974, 1985; Nagl *et al.*, 1985).

Occurrence of Polytene Chromosomes

Polytene nuclei were first observed in the larval salivary glands of Chironomidae, by Balbiani in 1881, but only at the beginning of the 1930s did Heitz and Bauer & Painter, independently and simultaneously, rediscover these enormous nuclei in the Malpighian tubules of *Bibio hortulanus* and in the larval salivary glands of *Drosophila melanogaster*, respectively (Ashburner, 1970). A few years later, Koltzoff, in 1934, and Bauer, in 1935, proposed the term polytenics for the giant chromosomes observed in these nuclei (Ashburner, 1970); polytene cells have since been described in many species (Nagl, 1978; Brodsky and Uryvaeva, 1985; Carvalheira and Guerra, 1998).

In plants, the first giant nuclei were observed by Osterwalder in 1898, in the enormous antipodal cells (antipodes) of the embryo sac of *Aconitum* (Nagl, 1981). However, as with the discovery of the giant cells of Chironomidae, the antipodal nuclei were largely forgotten for about 60 years. Only in 1956 did Tschermack-Woess and collaborators, during a reappraisal of the genus *Aconitum* genus and other plant species, recognize that the chromosomes observed in the antipodes were polytenics (Nagl, 1981). Unlike *Drosophila* polytene chromosomes, which present numerous bands and interbands, plant polytenics have a granular and fibril structure with no distinct bands. This structure probably occurs because of the absence of intimate synapsis between the sister chromatids. It is also believed that the chromocenter dispersion phase (Z phase) has some influence on the morphology of plant polytenics, as it results in a slight separation of these chromatids (Nagl, 1970a). However, Nagl (1969a) has reported that in *Phaseolus vulgaris* the structure of polytene chromosomes of embryo suspensor cells seems to be altered when these cells are submitted to low temperatures, becoming partly compacted and forming bands similar to those seen in *Drosophila*. Such results have not been observed again, remaining the only report of plant polytenics with bands and interbands.

Since the discovery of the polytene nuclei in antipodes, many other tissues composed of polytene cells have also been described. It is interesting to note that, until very recently, the cells with polytene chromosomes seemed to be limited to ovary tissues (antipodal cells, synergids, endosperm and embryo suspensor cells); however, polytenics have now been observed in anther hair, glandular hair and anther tapetal cells.

Polyteny can also be induced *in vitro* and it has been found that the meristematic tissues of root tips and the cotyledon cells of some plant species are able to form polytene chromosomes when submitted to specific treatments, including high temperatures (Shang and Wang, 1991) or an appropriate amount of certain growth regulators (mainly auxin and cytocinin) in the medium (Marks and Davies, 1979; Therman and Murashige, 1984).

Polytene Chromosomes of Embryo Suspensor

The most widely studied plant tissues with polytene cells is the *Phaseolus* embryo suspensor tissue. This tissue is found in the developing ovary of several angiosperm species (Esau, 1974). In *P. coccineus* and *P. vulgaris*, the suspensor is composed of about 200 cells, distributed between the basal and junction regions. The basal region is formed of about 20 giant mononucleate cells with a high level of polytenization, and with the DNA content of some cells being up to 8.192 C (Brady, 1973a,b). The junction region is composed of about 180 cells linking the basal region to the embryo proper. This last region possesses polyploid cells and/or cells with low polyteny level (Brady and Clutter, 1974).

The embryo suspensor provides nutritional support for the immature embryo, supplying proteins or synthesizing the substances necessary for embryo development (Schulz and Jensen, 1969). The underdevelopment of the cell wall of suspensor cells and their other structural characteristics indicate their secretory function in transporting nutrients through their membranes to the embryo (Nagl, 1974; Cionini, 1987). Analyses of the growth regulator level and transcription activity indicate that the suspensor tissue may play an important role during embryo ontogenesis, and seems to have a function in the synthesis of phytohormones needed for embryo development (Walbot *et al.*, 1972; Clutter *et al.*, 1974; Alpi *et al.*, 1975; Cionini *et al.*, 1976; Lorenzi *et al.*, 1978).

The basal cells of the embryo suspensor tissue of *P. coccineus* display 22 polytene chromosomes that are up to 30 times larger than the mitotic ones (Nagl, 1974), the 11 chromosome pairs having been identified earlier by their heteropycnosis pattern (Nagl, 1967). All mitotic chromosomes present heterochromatic centromeric bands and some weak interstitial and terminal ones (Schweizer and Ambros, 1979). Staining with the fluorochromes CMA and DAPI has revealed that most of these bands are CMA^+, although unlike mitotic bands they contain a small amount of $DAPI^+$ heterochromatin (Schweizer, 1976). The difference observed in the fluorescent pattern has been attributed to the better structural resolution of the giant chromosomes.

In situ hybridization experiments, with isotopic (q.v. Schumann *et al.*, 1990) and non-isotopic markers (q.v. Nenno *et al.*, 1994), have contributed considerably to the characterization of polytenics. These techniques have permitted both the location of many of their DNA sequences and the study of their replication cycle (Brady and Clutter, 1974).

For example, the cytolocalization of the ribosomal genes in *P. coccineus* has been demonstrated by the use of RNAr-H^3, revealing RNA puff activity both in satellite pairs and in the heterochromatin of chromosome pairs, without satellites (Avanzi *et al.*, 1971, 1972; Durante *et al.*, 1977). Isotopic techniques have also made it possible to observe the "extra nucleoli" associated with the telomeres of the polytenics without satellites, suggesting the existence of amplification in this region (Nagl, 1973). Actually, the extra DNA synthesis in the polytenics of the *Phaseolus* suspensor occurs at the beginning of embryogenesis and not simultaneously with the endoreduplication cycles (Avanzi *et al.*, 1970). Such gene amplification can occur both in the ribosomal cistrons and other regions of the genome and involves some polytene chromosome chromatids (Cremonini and Cionini, 1977). Some of these amplified regions are released from the polytenics to form micronucleoli. According to Avanzi *et al.* (1971), the micronucleoli are composed of a spherical mass of ribonucleoprotein covered by a layer of DNA. These micronucleoli seem to be associated with the intense metabolism of the suspensor basal cells (Nagl, 1973).

With advances in the fluorescence *in situ* hybridization technique (FISH), several other sequences have been located in plant polytene chromosomes. The first genic sequence hybridized in polytenics was that of the *phaseolin* group (Schumann *et al.*, 1990; Nenno *et al.*, 1993, 1994). This gene encodes the main seed storage protein of *Phaseolus* species. In these papers, it was demonstrated that the *phaseolin* gene seems to be located in chromosome 7 of *P. coccineus*. Another gene that has been located in the polytenics of *P. vulgaris* is the *PGIP* gene which encodes for polygalacturonase-inhibiting protein, a cell wall protein that specifically inhibits fungal endopolygalacturonases that are important during the early stages of plant pathogenesis. The *PGIP* gene has been located in a single region of the pericentromeric heterochromatin of the chromosome pair X, next to the euchromatin (Frediani *et al.*, 1993).

The FISH technique as applied in polytene chromosomes has also been a useful tool to study gene evolution. Nagl (1991) hybridized telomeric DNA and the *aromatase* gene sequence (both from human genome) in *P. coccineus* and *P. vulgaris* polytenics. The results showed that these sequences also hybridize with plant chromosomes, supporting the hypothesis of the evolutionary conservation of important coding or non-encoding sequences throughout living organisms.

Polytene Chromosomes of Anther Tapetum

Polytene chromosomes are also observed in other plant tissues, of which the anther tapetum tissue has made valuable contributions to the understanding of polytenics in angiosperms. This tissue is widely conserved, being found in groups ranging from bryophytes to angiosperms (Pacini and Franchi, 1993).

The tapetum is the innermost layer of the anther wall in close contact with the pollen grains. It is generally composed of a simple layer of cells, characterized by the presence of dense cytoplasm and quite a well-developed nucleus (Echlin, 1971). During the differentiation of the tapetum, the cells increase both their cytoplasmatic and nuclear volume and then undergo autolysis and degenerate (Mascarenhas, 1990). The tapetum's function seems to be related to the maturation of pollen grain, with biochemical and cytological studies demonstrating the intense metabolic activity in its cells at the end of the tetrad stage and during exine formation (Rowley, 1993).

Until recently, most of the information on the nuclear development of the tapetal cells has come from studies of anther ontogenesis, based on histological analyses by optical or electronic microscopy. Cytological analyses in tapetum were done mainly by Cooper (1933), Brown (1949), Oksala and Therman (1977), Franceschi and Horner (1979), and D'Amato (1984, 1989). However, cytogenetical analysis of the tapetal cells of *Vigna* species has revealed that this tissue can present very peculiar characteristics (Guerra and Carvalheira, 1994).

Anther tapetum cells are characterized by the presence of endomitotic or endoreduplication cycles (Cooper, 1933; D'Amato, 1984, 1989; Malallah *et al*, 1996). In species where the tapetum layer is composed of mononucleate cells, the increase in DNA content is generally a consequence of several endoreduplication cycles, while in species with bi- or multinucleate cells in the tapetum it is the endomitotic cycle which is responsible. In spite of the endoreduplication cycle producing mononucleate cells, tapetal binucleate cells with polytene nuclei have sometimes been observed (Carvalheira and Guerra, 1994).

In general, at the beginning of meiosis, the tapetal cells are mononucleate and diploid. During the tapetal differentiation, three cellular types can be observed, i.e. multinucleate cells, with more than one diploid nucleus (Malallah *et al.*, 1996), mononucleate cells, with a single polyploid nuclei (Carvalheira, G.M.G. and Guerra, M., unpublished data), and mono- or binucleate cells, with one or two polytene nuclei (Carvalheira and Guerra, 1994). In each of these cases, the DNA content per cell is often increased, suggesting that this tissue needs several copies of most genes to supply specific substances for exine development and consequent pollen grain maturation.

The increase in the ploidy level is probably caused either by suppression of anaphase movement (producing a "dumbbell"-shaped polyploid interphase nucleus) or by the occurrence of endomitotic cycles (D'Amato, 1989), while the increase in the number of nuclei per cell is due to the occurrence of one or more mitosis cycles without cytokinesis, resulting in multinucleate cells. Polytene nuclei, on the other hand, are formed through the endoreduplication cycles, and could remain in the interphase stage until the S phase, or progress to the prophase stage and return to a new G_1 phase (Guerra and Carvalheira, 1994).

Analysis of both ploidy level and nuclear structures in tapetal cells in genera of several subfamilies has revealed that their chromatin structure may be constant at the genus level. In the family Scrophulariaceae, for example, some species of the genera *Pedicularis* and *Melampyrum* have tapetal cells with tetraploid nuclei, in the post-meiotic period. On the other hand, in this same family, the genera *Odontetis*, *Euphrasia* and *Bellardia* have nuclei with enormous chromocenters, but with the same ploidy level (Greilhuber, 1974).

Like the other plant tissue with polytene chromosomes, anther tapetum cells can display polytenics that vary in structure and morphology from species to species (q.v. Nagl, 1974, 1981; Carvalheira and Guerra, 1994, 1998). For example, polytene nuclei in the antipodes of *Papaver rhoeas* were classified by Nagl (1981) into four different types, i.e., nuclei with chromocenters associated with radial chromatin bundles; decondensed nuclei with isolated chromatin fibers; nuclei with condensed chromatin, and nuclei with polytenics proper. Similar variation was found in suspensor cells of *Phaseolus* embryos, where the polytenics sometimes had granular or fibril form, depending on the degree of contraction in the interchromomeres (Nagl, 1978).

According to Carvalheira and Guerra (1998), the chromatin structure of the polytene nuclei in tapetal cells may basically be divided into three different types:

1. Individualized polytene chromosomes whose chromatin bundles are heteropycnotic in the proximal region and dispersed in the distal region, characteristic of *Phaseolus coccineus*, *P. vulgaris* (Carvalheira and Guerra, 1994), *Vigna unguiculata* (Guerra and Carvalheira, 1994; Carvalheira and Guerra; 1998), *V. umbellata*, *V. radiata* (Carvalheira and Guerra, 1998), *Lathyrus* and *Sesbania marginata*.
2. Polytene nuclei with chromocenters associated with the chromatin bundles, found in most of the species analyzed, including *Arachis hypogeae*, *Caesalpinea echinata*, *Clitoria cajanifolia*, *Crotalaria retusa*, in three *Habenaria* species, *Luffa cylindrica*, *Macroptilium peduncularis*, two *Phaseolus* species (Carvalheira and Guerra, 1994), *Pithecellobium dulce*, *Pisum sativum*, *Sophora tomentosa*, *Tropaeolum majus* and *Zomicarpa riedeliana*.
3. Polytene nuclei with chromocenters unassociated with chromatin bundles, a less frequent type, characteristic of *Genipa americana*, *Indigofera hirsuta*, *Lupinus polyphyllus*, *Vigna vexillata* (Carvalheira and Guerra, 1998) and *Vicia* sp.

Of these three types of chromatin organization, only the first is ideal for karyological analyses. The polytenics of this group is generally individualized and condensed. This type of chromatin organization has allowed good chromosome spreads to be obtained, facilitating the chromosome counting. In most *Phaseolus* and *Vigna* species analyzed that had this type of organization, it was possible to observe all 22 chromosomes of the karyotype (q.v. Guerra and Carvalheira, 1994; Carvalheira and Guerra, 1994).

Although this type of chromatin organization seems to be ideal for karyological analyses, these polytene chromosomes are somewhat smaller than those observed in the embryo suspensor of *Phaseolus* or those of some other genera. The largest polytenics of the anther tapetal cells (observed in *Vigna unguiculata*) are about 3.5 times bigger than the mitotic ones (Guerra and Carvalheira, 1994). The difference in size observed between polytenics of the embryo suspensor and anther tapetum is probably related to the number of endoreduplicated cells present in each of these tissues.

As was mentioned previously, only a few basal cells undergo many endoreduplication cycles in the embryo suspensor tissue (Nagl, 1981), while in the anther tapetum hundreds of cells undergo this process. The low level of chromatin endoreduplication associated with a large number of cells seems to satisfy the metabolic necessities of both the anther tapetum and microspores. On the other hand, in the giant cell of the embryo suspensor, many endoreduplication cycles seem to be necessary to maintain the perfect functional and nutritional stage of the embryo, most of whose nutrients are supplied by the suspensor cell.

Although the polytenics of the anther tapetum are reduced in size, they have helped in the cytolocalization of DNA sequences. The first *in situ* hybridization in tapetal polytenics has revealed interesting data different from that which was observed in the polytene chromosomes of the embryo suspensor (Guerra and Kenton, 1996). As stated earlier, after *in situ* hybridization with the human telomere DNA probe, Nagl (1991) observed that the embryo suspensor polytenics of *Phaseolus* show a group of dots or compact bands at the telomeres. However, when synthetic telomere oligomers were hybridized with tapetum polytenics in an amphidiploid hybrid of *Phaseolus*, the oligo was preferentially located at, or close to, the chromocenters.

These fluorescent areas were distributed randomly in the nuclear area, although association with the nuclear boundary was never observed (Guerra and Kenton, 1996). This may suggest that, at least in some aspects, the basic molecular organization of diploid nuclei in the anther tapetum is not completely conserved after the endoreduplication cycles.

In fact, the loss of telomere association with the nuclear membrane has been documented in some special chromosome types, such as pigeon lampbrush chromosomes (Solovei and Macgregor, 1995) and Diptera polytenic chromosomes (Agar and Sedat, 1983). In plant polytene nuclei, however, this loss of telomeric association with the nuclear envelope was reported for the first time in anther tapetum polytenics (Guerra and Kenton, 1996).

These chromosomes have also helped in the identification of the 45S ribosomal sites in *Phaseolus coccineus* and *Vigna unguiculata* (Guerra *et al.*, 1996), and 5S sites, in *V. radiata* and *V. unguiculata* (Carvalheira *et al.*, 1998). In *P. coccineus*, six ribosomal sites were observed in tapetal cells, as has been reported previously (see Avanzi *et al.*, 1972; Durante *et al.*, 1977). Surprisingly, however, ten ribosomal sites were observed in *V. unguiculata* tapetum polytenics (Guerra *et al.*, 1996), instead of one or two pairs as earlier reported (Frahm-Leliveld, 1965; Barone and Saccardo, 1990; Galasso *et al.*, 1992). The large number of ribosomal sites observed in *V. unguiculata* when compared with *Phaseolus* has suggested that this increase in ribosomal sites may have been initiated by genetic mechanisms, such as gene conversion.

According to Guerra *et al.* (1996), the variation in the number of rDNA sites observed between species of related taxa could be due to the differential amplification and fixation of rDNA sequences at different chromosomal sites. On the other hand, four 5S ribosomal sites were observed in *V. radiata* and *V. unguiculata* (Carvalheira *et al.*, 1998), confirming the previous reports for *V. unguiculata* (Galasso *et al.*, 1995), although these reports were first published for *V. radiata.*

In conclusion, although the polytenics of the anther tapetum are smaller than those in suspensor cells, both the large number of polytene cells in this tissue and their structural polytene morphology make these chromosomes more convenient for the study of plant polyteny and chromosome organization. Guerra and Carvalheira (1994) and Carvalheira and Guerra (1994, 1998) have suggested that such chromosomes present cycles of diffuse and condensed stages. The change from diffuse to condensed stage seems to depend on the endoreduplication level, genetic background and environmental factors. All these observations suggest that bundled polytene chromosomes of plants, at least in tapetal cells, are most probably the consequence of advanced endoreduplication cycles resulting in prophase or prophase-like chromosomes that may still be able to perform some DNA and RNA synthesis (Brady and Clutter, 1974; Cionini *et al*, 1982).

7

CYTOGENETIC TECHNIQUES

CHROMOSOME BANDING

A. Local variations in chromatin coiling, and the consequent variations in chromatin density, result in distinct staining patterns that can be used to identify each chromosome. The discovery of techniques which produce banding patterns on chromosomes is one of the most significant developments in cytogenetics. These techniques told us that chromosomes have an identity and structure that is consistent in all cells within a species, supporting the chromosome theory of inheritance. Staining made it possible to construct the first chromosome maps, and to distinguish between chromosomes of many different species. Fixed chromosomes are not equivalent to chromosomes in vivo. Fixation does not appear to extract DNA from chromosomes but it does extract histones and nonhistone proteins in varying amounts. The fixative may not always have a quantitative, reproducible effect on these proteins. The net result is that when you look at fixed chromosomes, you are looking at chromosomes that are chemically and structurally different from chromosomes in living cells.

Some relevant definitions:

- band - part of a chromosome that is clearly distinguishable from its adjacent segments by appearing lighter or darker.
- chromomere - any part of a chromosome that stains uniquely compared most bands.

- chromatin - the DNA/protein complex making up chromosomes
- heterochromatin - chromatin that tends to take up large amounts of stain, thus staining darkly. Heterochromatin tends to be densely-coiled.
- euchromatin - chromatin that tends to take up small amounts of stain, thus staining lightly.
- constitutive heterochromatin - chromatin that remains densely-coiled through out interphase, in all tissues. Often found near centromeres, telomeres and nucleolar organizing regions (NORs).
- facultative heterochromatin - chromatin that is either densely-coiled or loosely coiled during interphase, depending of the developmental state or tissue.

Giemsa Staining Methods

Mitotic metaphase is the best stage for studying chromosome morphology. Giemsa stain is specific for constitutive heterochromatin adjacent to centromeres and telomeres.

1. Geimsa C-banding technique has been used to identify individual chromosomes in many species by showing the position of constitutive heterochromatin.

Geimsa staining methods were first described in a paper by Purdue and Gall (1970). However, it is important to recognize that staining methods typically need to be fine-tuned for each species, to give the optimal number of bands.We will concentrate on the general treatments and the basis for their application in terms of what each treatment is doing to produce the banding pattern.

Steps of C-banding:

1. Roots are harvested, pretreated and fixed in 3:1 95% ethanol:glacial acetic acid for at least 24 h. The roots are softened in 45% acetic acid or in 0.5% aceto-carmine. Slides are prepared by the squash method and the cover glass is

removed using dry-ice method. Chromosomes adhere to the slide surface.

The acid pretreatment depurinates the DNA (ie. removes purine bases without cleaving the sugar-phosphate backbone. This helps to induce nicks in the DNA backbone, loostening any supercoiling.

2. Dehydration. Typically slides are placed in 95 to 100% ethanol for 1 hour.
3. Denaturation. Treatment with barium hydroxide for 5 to 15 min at elevated temperature 50-55° C to denature the DNA. Alkali treatment denatures the DNA. ie. breaks base pairing, resulting in ssDNA.
4. Renaturation. The slides are then washed with distilled water and transferred to incubation at 60° C in saline sodium-citrate solution SSC (NaCl) which presumably renatures the DNA. Incubation periods and temperature are variable.

   The incubation in SSC alters the composition of the chromatin (DNA- protein complex). The differential reannealing of the C-band and the non-C-band chromatin to extraction is the basis for the differential staining. The heterochromatin which is highly repetitive DNA renatures under these conditions while the middle repetitive DNA and unique DNA does not. The result is the C-banding pattern.
5. Staining. Slides are then stained with Geimsa stain and checked periodically to see how the stain is progressing. When the optimal staining has been achieved, the slides are rinsed in distilled water to remove the excess stain, air-dried, stored in xylene overnight, air dried again and the cover slip is mounted using Canada Balsam, etc.

Other Geimsa banding techniques include G banding which provides more detail than C-banding. G banding provides identification of almost every chromosome in a complement and structural variation can be detected.

G-banding is produced using Giemsa (=G) staining and usually pretreatment with a diluted trypsin solution, urea or protease.

The Human Chromosome Study Group has published a diagrammatic representation of chromosome bands observed with G-,Q-, and R-staining methods. The nature of these bands appears to be stronger chromosome condensation, but the bands could also be interpreted as the result of alteration of histones and other proteins of the chromosomes. G bands cannot be produced in plant chromosomes possibly because of the high degree of condensation in plant metaphase chromosomes. N-banding was originally developed to stain the nucleolar organiser region of the chromosomes, plant and mammals.

STRUCTURAL BASIS OF CHROMOSOME BANDING

The exact role of underlying structural, functional and molecular organization of the chromosome in the determination of bands is still not fully understood. C-bands represent constitutive heterochromatin. G-banding patterns on mitotic chromosomes correspond very closely to the chromomere patterns of meiotic chromosome bivalents at pachytene, that is, the chromatin that is densely-packed enough to see. Chromomeres appear to be the focus of chromatin condensation along the chromosome, and may be the sites where chromatin condensation is initiated. If this is true, G-bands may be the corresponding initiation sites for chromosome condensation during mitosis.

Chromomeres may be more apparent in meiosis because of the relative degree of condensation of the chromosomes. Meiotic chromosomes are very extended and the homologues are paired which enhances chromomere patterns. Mitotic chromosomes are overall more condensed, masking chromomeres and requiring banding pretreatment to reveal them.

On human mitotic prophase chromosomes, up to 2000 or more bands can be observed. The bands on these high-resolution chromosomes correspond to meiosis chromomeres. As the chromosomes condense in the prophase to metaphase transition, there is a progressive coalescence of bands such that each band and its component sub-bands retain the same relative location and staining intensity. The bands observed at metaphase are made up of a collection of smaller sub-bands.

The amount of constitutive heterochromatin in each chromosome is characteristic but polymorphisms do occur. The banding patterns are consistent and reproducible. Generally, features of chromosomes do not vary from tissue to tissue or change during the course of development. The organization reflected in banding is not random. In other words, we see consistent bands because the chromosome coils exactly the same way every time.

Constitutive heterochromatin is naturally differentiated by darker appearance during interphase and prophase, but does not usually show up in metaphase. It takes special methods to produce the bands.

In plant mitotic chromosomes, C-banding has been produced, but not G-banding , although pachytene chromosomes of plant meiotic cells demonstrate chromomeres. One suggestion is that this is due to the much greater quantity of DNA per unit length in plant chromosomes compared to those in mammals. The chromomeres are too tightly packed to be resolved by G-banding. The reason is not clear as compaction seems to be roughly equivalent. The preparation or pretreatment may have to be adjusted.

POLYTENE CHROMOSOMES ALLOW BANDING AT A VERY FINE LEVEL OF RESOLUTION

In specialized cells of some plant and animal species, chromosomes undergo many cycles of DNA replication without undergoing mitosis. The result is polytene chromosomes, which may contain as many as a thousand parallel copies of each chromosome, perfectly aligned along their entire length. Polytene chromosomes can be stained and viewed in interphase nuclei. The most common example is the polytene chromosomes in the salivary glands of

Drosophila. Because interphase chromosomes are more extended than those in mitotic cells, loci can be identified on polytene chromosomes at a much finer level of resolution.

Fluorescence In-situ Hybridization

Fluorescence In-situ hybridization (FISH) uses fluorescence to highlight the site of specific DNA sequences. Trask et al. 1993 provide a detailed review of the technique and its applications. In essence, a DNA sequence is 'tagged' by incorporating fluorescently-labeled nucleotides into the DNA chain. The tagged DNA, referred to as a probe , is incubated with chromosomes on a slide. Wherever the chromosome contains the same sequence as the probe, the probe can base-pair with the chromosome. The site to which the probe base pairs will fluoresce in UV microscopy.

In-situ hybridization is a valuable technique for studying chromosomal distribution of specific nucleic acid sequences and to determine the location of highly repetitive DNA sequences. Changes in these highly repetitive sequences reflect the physical rearrangements in the genome over time in the evolution of the species. Highly repetitive DNA sequences represent between 20 and 90% of the DNA in higher plants. The rate of evolution of individual repetitive sequences varies considerably; some may be conserved and be present in many species within a taxonomic family. The original technique published by Gall and Pardue (1969) has undergone numerous modifications and fluorescent in-situ hybridization (FISH) has been applied to chromosome identification and diagnosis in human genetic studies.

As with Geimsa banding, chromosomes must be fixed before denaturation. It is necessary to denature both the DNA probe and the target chromosome to allow complementary pairing to take place.

Fixative: The chromosomes are fixed in a 3:1 methanol:acetic acid solution. Slides are prepared by the squash method and the cover glass is removed to allow the cells to dry. The squash method stabilizes the chromosomes, permits good penetration of the reagents and brings most of the chromosome into the same focal plane, and also increases the speed at which the hybridization takes place, by decreasing volume.

Denaturation : incubate at 72° C in 50-70% formamide and 2X SSC (NaCl) (formamide denatures DNA helices)

Addition of Probes: The probe molecules are pieces of DNA, selected to target specific sequences on the chromosome. The target sequences can be as small as 1kbp or larger than 15 kbp. Sequences can be taken from entire chromosomes to highlight large regions of the genome. Denatured probe is hybridized with fixed chromosomes under a coverslip overnight at 37° C in 50% formamide, 2X SSC and 10% detran sulfate. After incubation, unannealed probe is washed off and the slides are labelled with immuno-fluorescent tags (ie. antibodies to the labeled nucleotides carrying fluorescent tags) and an anti-fading agent.

In summary, the chromosomes have been exposed to methanol, acetic acid, flattening, drying, rehydration, heat, formamide, and high salt concentrations during the FISH procedure. Some degree of structural changes will have been produced in the chromosomes. High resolution microscopic equipment is required for FISH as well as image processing software to enhance and analyse the image.

FISH is being used to study chromosome banding patterns, the distribution of chromatin in interphase nuclei, and chromosome abnormalities. Greater resolution is possible in interphase chromosomes due to the dispersed state of the chromatin, so that small structural changes can be detected. The compaction of metaphase chromosomes sets the limit of detection at sequences separated by more than 1 Mbp. The limit in interphase chromosomes is a separation of 100 kbp (Figure 8, Trask et al. 1993).

Detection of Single Genes

When the probe is a single gene, only that gene is expected to light up in FISH. Figure 1 shows hybridization of a probe for the *neu* proto-oncogene in hyman lymphocytes. The *neu* locus resides on chromosome 17. Since the cells contain diploid nuclei, two hybridizing spots are seen in each nucleus. The sensitivity of FISH is put into perspecive by considering the fact that each chromosome is a single DNA molecule. That is, each spot results from a few probe molecules hybridizing to different parts of a single DNA molecule. Looked at another way, we can recall that Avogadro's number (the number of molecules in a mole) is 6.02×10^{23} molecules/ mole, we are detecting 1.66×10^{-24} moles of chromosomal DNA!

Chromosome Painting

It is possible to isolate DNA specifically from individual chromosomes. When total chromosomal DNA is used as a probe, it will base pair with all sequences on the chromosome from which it was derived, making the entire chromosome 'light up'. In other words, each sequence in the probe eventually encounters its complementary sequence on the chromosome and base-pairs with it.

8

CELL AND TISSUE CULTURE AND MUTATION INDUCTION

BANANA CELL AND TISSUE CULTURE

The International *Musa* germplasm collection is sited at the INIBAP (International Network for the Improvement of Banana and Plantain) Transit Centre at K.U.Leuven. By now, more than 1000 different accessions of shoot-tip cultures have been initiated *in vitro*, multiplied and maintained at reduced temperature conditions (16 ± 1°C). Shoot cultures are grown on MS (Murashige and Skoog) medium, supplemented with 30 g/l sucrose, 2.25 mg/l BA (6-benzyladenine) and 0.175 mg/l IAA (indole-3-acetic acid). In comparison with the culture medium on which shoot-tips are maintained, a tenfold decrease in cytokinin content (0.225 mg/l BA) induces regeneration of rooted plants. In contrast, adding 22.5 mg/l BA to the culture medium results in suppression of the apical dominance in shoot-tip cultures and a reduction of corm and leaf tissue between meristematic tissue.

Highly proliferating meristem cultures are obtained and used as starting material in the scalp methodology, the technique most commonly applied for the development of embryogenic cell suspensions at the Laboratory of Tropical Crop Improvement, K.U.Leuven. Initiation and maintenance of cell cultures is rather labour intensive and time consuming. However, since 1 ml of settled cells of a highly regenerable cell suspension can yield more than 100,000 plants, cell cultures are most suitable for mass clonal propagation. Moreover, embryogenic cell suspensions are highly preferred as target material for protoplast culture and genetic engineering since the risk of chimerism is circumvented because of the unicellular origin of regenerated plants.

Plant tissue culture is the science of growing plant cells, tissues or organs isolated from the mother plant, on artificial media. It includes techniques and methods appropriate to research into many botanical disciplines and several practical objectives. Both organized and unorganized growth is possible *in vitro*.

Organized growth of banana tissue *in vitro* is limited to embryo culture and shoot tip culture. Embryo culture is an important aid for classical breeding in banana since the germination frequency of seed is extremely low. Through embryo rescue, this frequency can be 10 times increased.

Major applications of shoot tip culture are mass clonal propagation and germplasm conservation. In the former already existing shoot tips are stimulated to multiply rapidly, while in the latter, the multiplication rate is slowed down. This technique is applied for the conservation of The International *Musa* germplasm collection at the INIBAP (International Network for the Improvement of Banana and Plantain) Transit Centre at K.U. Leuven.

In vitro culture of unorganized tissue in banana is almost exclusively related to the establishment of embryogenic cell cultures. When an embryogenic callus is induced on solid media containing high auxin concentrations, it can be transferred to liquid medium where it gives rise to embryogenic cell suspensions. Such suspensions with high regeneration capacity can be used for mass clonal propagation and are the only source of regenerable protoplasts in banana. More importantly, they are the preferred target material for induced mutations and genetic engineering.

In this article, a brief overview is given on the most common techniques in banana cell and tissue culture and their applications.

SHOOT-TIP CULTURES

Micropropagation

Stage 1: Initiation of shoot cultures

Shoot cultures of banana start conventionally from any plant part that contains a shoot meristem, i.e. the parental pseudostem, small suckers, peepers and lateral buds. The apex of the inflorescence and axillary flower buds are also suitable explants for tissue culture initiation. Overall, it is important to select explant material from preferably mature individuals whose response to environmental factors is known, and whose quality traits governed by genotypic and environmental effects have been identified.

For rapid *in vitro* multiplication of banana, shoot tips from young suckers of 40-100 cm height are most commonly used as explants. From the selected sucker a cube of tissue of about 1-2 cm^3 containing the apical meristem is excised. This block of tissue is dipped in 70% ethanol for 10 s, surface sterilized in a 2% sodium hypochlorite solution, and after 20 min rinsed three times for 10 min in sterile water. Variants of this decontamination protocol exist. They differ in explant type and size, disinfection procedure (single or double sterilisation), type of disinfectant (calcium hypochlorite instead of sodium hypochlorite) and its concentration and treatment duration. Subsequently a shoot tip of about 3 × 5 mm, consisting of the apical dome covered with several leaf primordia and a thin layer of corm tissue, is aseptically dissected. Larger explants have the merit of consisting of a shoot apex bearing more lateral buds which rapidly develop into shoots.

The optimal size of the explant depends on the purpose. For rapid multiplication, a relatively larger explant (3-10 mm) is desirable despite its higher susceptibility to blackening and contamination. When virus or bacteria elimination is needed, meristem-tip culture is the preferred option. The explant is then further reduced in size (0.5-1 mm length), leaving a meristematic dome with one or two leaf initials. Meristem cultures have the disadvantage that they may have a higher mortality rate and an initial slower growth.

The explant is placed directly on a multiplication-inducing culture medium. For banana micropropagation, MS-based media are widely adopted. Generally, they are supplemented with sucrose as a carbon source at a concentration of 30-40 g/l. Banana tissue cultures often suffer from excessive blackening caused by oxidation of polyphenolic compounds released from wounded tissues. These undesirable exudates form a barrier round the tissue, preventing nutrient uptake and hindering growth. Therefore, during the first 4-6 weeks, fresh shoot-tips are transferred to new medium every 1-2 weeks. Alternatively, freshly initiated cultures can be kept in complete darkness for one week. Antioxidants, such as ascorbic acid or citric acid in concentrations ranging from 10-150 mg/l, are added to the growth medium to reduce blackening, or the explants are dipped in antioxidant solution (cysteine 50 mg/l) prior to their transfer to culture medium.

Usually two types of growth regulators, a cytokinin and an auxin, are added to the banana growth medium. Their concentration and ratio determines the growth and morphogenesis of the banana tissue. We routinely add 2.25 mg/l 6-benzyladenine (BA) and 0.175 mg/l indole-3-acetic acid (IAA) to the initiation medium.

In most banana micropropagation systems, semi-solid media are used. As a gelling agent agar (5-8 g/l) is frequently added to the culture medium but our preference is for Gelrite (2-4 g/l) because of its higher transparency, allowing much earlier detection of microbial contamination. Liquid media are superior for shoot multiplication , but for maximum plant production and survival *ex vitro*, one culture cycle on semi-solid medium is also needed.

Banana shoot-tip cultures are incubated at an optimal growth temperature of 28 ± 2°C in a light cycle of 12-16 h with a photosynthetic photon flux (PPF) of about 60 $\mu E/m^2s^1$.

Stage 2: Multiplication of shoot-tip cultures

The formation of multiple shoots and buds is promoted by supplementing the medium with relatively high concentrations of cytokinins. In banana, BA is the preferred cytokinin and is usually added in a concentration of 0.1-20 mg/l. For the multiplication of propagules, we use the same medium as for the initiation of shoot cultures (p5 medium containing 2.25 mg/l BA and 0.175 mg/l IAA). If the production of highly proliferating meristem cultures is required (Section 3.2.1.), a tenfold higher concentration of BA is added to the culture medium (p4 medium containing 22.5 mg/l BA and 0.175 mg/l IAA). Higher concentrations of the cytokinin BA tend to have an adverse effect on the multiplication rate and morphology of the culture and should therefore be avoided. The rate of multiplication depends both on the cytokinin concentration and the genotype. In general, shoot tips of cultivars having only A genomes produce 2-4 new shoots, whereas cultivars having one or two B genomes produce a cluster of many shoots and buds at each subculture cycle. Approximately 6-12 weeks after culture initiation, depending on the initial explant size, new axillary and adventitious shoots may arise directly from the shoot-tip explant. Clusters can be separated, trimmed and repeatedly subcultured at 4-6 week intervals.

Stage 3: Regeneration of plants

Individual shoot or shoot clumps are transferred to a nutrient medium which does not promote further shoot proliferation but stimulates root formation. The cytokinin in the regeneration medium is greatly reduced or even completely omitted. Within 2 weeks, shoot tips develop into unrooted shoots. To initiate rhizogenesis IAA, NAA (a-naphthalene acetic acid) or IBA (indole-3-butyric acid) are commonly included in the medium at between 0.1 and 2 mg/l. We use the same auxin concentration as in the proliferation medium (0.175 mg/l IAA), but a tenfold lower BA concentration (0.225 mg/l). For some genotypes (*Musa* spp. ABB and BB group) that produce compact proliferating masses of buds, activated charcoal (0.1-0.25%) is added to the regeneration/rooting medium to enhance shoot elongation and rooting. After rooting, plants are hardened *in vitro* for 2-4 extra weeks on the regeneration/rooting medium prior to transplantation to soil.

Conservation of Shoot Cultures

Shoot or meristem tip cultures are suitable not only for the large-scale production of uniform and vigorously growing propagules for field establishment, but also for germplasm conservation. For banana species which are predominately propagated vegetatively, *in vitro* techniques are thus complementary to field conservation.

Banana species can be stored under normal growth conditions at 28°C, but this involves transfer every 2-4 months. Hence there is the risk of losing material through microbial contamination or through human error (mislabelling of culture vessels). Nevertheless, conservation of germplasm under short-term storage conditions is most useful, for instance, to breeders or researchers who wish to maintain a small working collection of breeding material during evaluation, testing, selection and hybridisation, or to maintain products of *in vitro* manipulation.

Slow growth conservation has the benefit of reducing the number of subcultures, thus making significant savings in labour input. Germplasm still remains readily available for regeneration, multiplication and distribution. Minimal growth is clearly useful for conservation of genotypes, but is constrained by the risk of genetic changes (somaclonal variation) resulting in the loss of distinct genotypes.

Suppression of growth can be achieved by various modifications of the physical and/or chemical tissue-culture environment. The most common and widely applied growth-retarding factor in banana is low temperature. This technique has been used routinely for many years for the conservation of the International *Musa* Germplasm collection at the INIBAP (International Network for the Improvement of Banana and Plantain) Transit Centre at K.U.Leuven. Banana shoot cultures are stored at 16 ± 1°C. Each accession in the collection is represented by a set of 20 replicates of shoot cultures grown on MS medium, supplemented with 30 g/l sucrose, 2.25 mg/l BA and 0.175 mg/l IAA. Subculture intervals are extended to about 12 months, although large differences in storage potential have been observed among the different genotypes: some genotypes keep for about 20 months whereas others require subculturing every 2 months.

Alternatively, the use of osmotica in banana tissue cultures has been investigated. The addition of 4% mannitol and 3-6% sucrose to the growth medium resulted in 50% growth reduction. Cultures grown on mannitol-enriched medium at 27°C could be kept for 6 months, whereas cultures on the control medium (3% sucrose) started to deteriorate after 4 months of incubation.

Contamination

Contamination in tissue cultures may be caused by endogenous bacteria that escape initial disinfection or by micro-organisms introduced during tissue-culture manipulations. Both types of contaminants may survive in the plant material for several subculture cycles and over extended periods of time without expressing symptoms in the tissue or visible signs in the medium.

'Internal' bacteria are a considerable source of concern in all aspects of plant cell, tissue and organ culture because they hinder the international exchange of germplasm or become a nuisance when contaminated tissues are used as explant material for cryopreservation or for the initiation of embryogenic cell suspensions. It is therefore important that control measures are taken at every tissue-culture step. At the INIBAP Transit Centre, plant material is tested for endophytic bacteria on a broad spectrum bacteriological medium at tissue culture initiation and during annual subculturing. Prior to placing each shoot-tip on culture medium, the base of the explant is streaked onto Difco Bacto nutrient agar, enriched with glucose (1%) and yeast extract (0.5%) in Petri dishes. Testing reveals the presence of cryptic contaminants in 5% of the stored germplasm. In mass propagation systems, positive stock materials should immediately be destroyed. However, germplasm in collections or cultures that are not readily replaceable with fresh material (e.g. from field or from greenhouse stock plants) are 'cleaned up'.

Some antibiotics are successful in controlling bacterial contaminants in banana tissue cultures. Rifampicin (100 mg/l) added to liquid cultures for 10-30 days was found most suitable for controlling frequently occurring Gram-positive bacteria in banana shoot tips without affecting plant growth.

However, culturing small meristem tips (1 mm) isolated from contaminated *in vitro* plants or from greenhouse plants obtained from contaminated *in vitro* plants was found to eliminate any bacterial contaminant.

Somaclonal Variation

Shoot-tip culture preserves genetic stability much better than callus or cell suspension cultures, yet somaclonal variation appears to be widespread among plants regenerated from banana shoot-tip cultures.

Off-type frequencies vary from 1% to 74%. The phenomenon of somaclonal variation in the plantain subgroup (*Musa* spp. AAB group) was extensively studied.

It revealed that the incidence of somaclonal variation is strongly influenced by the genetic stability of each cultivar, and that its frequency is amplified by culture-induced factors.

There is no evidence that growth regulators routinely used in tissue culture directly affect the rate of variation, but it has been found that the rate of somaclonal variation is positively related to the generation number.

For 'Williams' (AAA) *in vitro* plants obtained after one and five subculture cycles, dwarfism and leaf-off types counted for 3.7% and 0.7% respectively after one *in vitro* cycle and increased to 6.1% and 1.9% respectively after five *in vitro* cycles.

It is therefore recommended that the number of subculture cycles should be limited to ten or that the number of plants produced from a primary explant should be limited to no more than 1000.

SOMATIC EMBRYOGENESIS

Procedures to Produce Embryogenic Cell Suspensions: an Overview

In bananas and plantains, there are four main procedures for the development of embryogenic cell suspensions. They differ mainly by the explant in which embryogenesis is induced: zygotic embryos, rhizome slices and leaf sheaths, immature (fe)male flowers, and proliferating meristem cultures.

Each method has its own limitations, rendering the establishment of *Musa* embryogenic cell suspensions still far from routine. Since most edible banana cultivars rarely set seeds, zygotic embryos are of limited value as starting material. Reports on the efficiency of embryogenesis induction in highly differentiated rhizome and leaf tissue are scarce.

The scalp-methodology (described below) relies on proliferating meristem cultures as explants. This involves an extensive material preparation phase preceding induction of embryogenesis.

In contrast, the starting material for the widely used male-flower technology can be collected directly from the flowering banana plants. A fast decline of the male flower embryogenic response soon after harvest, as well a seasonal dependence, requires direct field access and inoculation of explants quickly after harvest.

In addition, this method cannot be applied to the highly preferred False Horn and Harton plantains, which do not produce male flowers. This can be overcome by the use of female flowers as alternative explants, but results in loss of the bunch and is therefore difficult for large-scale use.

Scalp-derived Embryogenic Cell Suspensions

At the Laboratory of Tropical Crop Improvement, K.U.Leuven, research on somatic embryogenesis is focused on the scalp methodology.

Preparation of embryogenesis competent explants (scalps)

Embryogenesis competent scalps are 3-5 mm explants containing a high number of tiny white meristems with a small amount of corm or leaf tissue.

The time needed to prepare highly proliferating meristem cultures from which scalps of good quality can be excised ranges from a few months to more than one year.

For cultivars like 'Bluggoe' (ABB group) with a high initial *in vitro* proliferation rate, suitable starting material can be obtained on standard proliferation medium (p5 medium with 2.25 mg/l BA). For other types like Plantains (AAB group), Cavendish (AAA group) and East African Highland bananas (E-AAA group), several cultures on medium enriched in cytokinin (p4 medium with 22.5 mg/l BA) are needed.

Table: Different* in vitro *phases involved, behaviour of plant material and duration of different steps in the establishment of scalp-derived embryogenic cell suspensions.

Phase	Behaviour of plant material	Duration (months*)
1. Preparation of embryogenesis competent explants (scalps)	Homogeneous proliferation; good quality scalps	5-14
2. Embryogenesis induction	Embryogenic complexes (globules, embryogenic cells, embryos)	4-7
3. Suspension initiation and upgrading	Embryogenic cell suspensions	3-6
	Subtotal	12-27
4. Regeneration	Rooted plantlets of test tube size	3-8
	Grand total	15-35

* Genotype and cultivar dependent

Proliferating meristem cultures that allow scalp preparation can be obtained in theory for any landrace. However, the extent that corm and leaf tissue can be reduced between the meristematic tissue is dependent on the genomic constitution and even the cultivar. The minimum number of cycles on p4 proliferation medium was negatively correlated with the percentage of B chromosome sets in the genome.

Recently a broad range of cytokinins [BA, thidiazuron (TDZ), 6-g,g-(dimethylallylamino) purine, zeatin and kinetin] was explored for their influence on shoot tips freshly excised from *in vitro* rooted plantlets. Based on percentages of outgrowth and multiplication, and the amount and length of developing shoots, the cytokinins can be ranged as follows (from strongest to weakest activity in triggering multiplication): TDZ > BA > kinetin > zeatin > 2iP.

The lengthy material preparation phase for highly proliferating meristem cultures could be reduced up to threefold by (i) inoculation of freshly excised 5 mm explants (instead of shoot tip cultures) and (ii) the use of TDZ (instead of BAP) as cytokinin. Investigations are currently being conducted to find out whether these highly proliferating TDZ cultures are also competent for embryogenesis.

Induction of somatic embryogenesis

Embryogenesis is induced by scalp inoculation onto semi-solid medium containing 1 mg/l 2,4-D (2,4-dichlorophenoxyacetic acid) and 0.22 mg/l zeatin as plant growth regulators (ZZ medium). According to the embryogenic response (no embryogenic response, occurrence of individual embryos, presence of embryogenic callus), three main patterns of development are found.

The formation of fast-growing, yellowish white callus during the first weeks after embryogenesis induction is not wanted since such calluses eventually become necrotic, most often without any embryogenic structure. A positive embryogenic response occurs generally on 3-8-month-old induced explants.

The appearance of individual embryos is a promising indication of the embryogenic capacity of the starting material. More interesting is a white callus, consisting of only early-stage somatic embryos and non-organised embryogenic cell clusters. Only these friable embryogenic complexes are suitable for the initiation of embryogenic cell suspensions. The nature of the embryogenic response depends not only on the genotype but also on the selected line and even the experiment.

Table: Genomic constitution, variety, type and corresponding frequency of embryogenic callus encountered on scalps induced for embryogenesis

Genomic constitution	Variety	Type	Successful induction of embryogenesis (%)*
AA	Calcutta4	Wild diploid	0
AAA	GN FHIA	Cavendish	0-2.9
AAA	GN JD	Cavendish	0-4.2
AAA	Gran enano	Cavendish	0-11.7
AAA	Williams BSJ	Cavendish	0
AAA	Williams JD	Cavendish	0-22.2
E-AAA	Ingarama	Highland	0
E-AAA	Mbwazirume	Highland	0
E-AAA	Nyamwihogora	Highland	0
AAB	Agbagba	Plantain	0-0.5
AAB	Obino l'Ewai	Plantain	0-2
AAB	Orishele	Plantain	0-5.8
ABB	Burro Cemsa	Cooking banana	0

* lowest and highest frequency (%) of embryogenic callus encountered in a single experiment

Initiation and maintenance of embryogenic cell suspensions

Success rates for the initiation of embryogenic cell suspensions depend largely on the quality and volume of available embryogenic complexes. In our experience, it is not worthwhile to transfer distinct embryos to liquid ZZ medium.

Overdeveloped embryos either turn black due to oxidation of phenolic compounds or dedifferentiate into globules which only release non-embryogenic cells. In contrast, homogeneous complexes consisting of a high proportion of embryogenic callus and early-stage transparent embryos are suitable as inoculum. The first few months following the initiation of embryogenic cell suspensions are labour-intensive.

This is mainly due to the heterogeneity of the freshly initiated cultures. Components of young cell suspensions and their evolution in time are discussed by Schoofs. To avoid differentiation and to stimulate multiplication of embryogenic cell clusters, the maintenance medium must be refreshed weekly. In addition, to maintain the regeneration capacity of the cell suspension, globules which release only starchy dense or empty cells have to be discarded.

On average six months after initiation, embryogenic cell suspensions reach the phase of mass multiplication. The maintenance medium of the cell cultures is then refreshed every 2 weeks with an optimal initial inoculum density ranging from 1.5 to 3%.

A two- to threefold increase in settled cell volume is reached at the end of each subculture period. Even at this level, embryogenic cell suspensions remain more or less heterogeneous.

Plant regeneration from embryogenic cell suspensions

Regeneration from banana cell suspensions is not really a problem when the suspension consists entirely of embryogenic cell clusters. Medium composition, light conditions and inoculum age only slightly affect the regeneration capacity.

Based on weight measurements and counting of germinating embryos and plants, the regeneration capacity of cell suspensions established using the scalp methodology ranges between 10^4 to 10^5 somatic embryos per millilitre of settled cell volume. The conversion rate of germinating embryos into rooted plantlets is 90-100%.

These results are in accordance with data obtained for male flower-derived embryogenic cell suspensions. The number of plants obtained from 1 ml settled cells of several scalp-derived cell suspensions is given in Table 3. Assuming (i) an initial inoculum density of 1.5% settled cell volume in 60 ml ZZ maintenance medium, and (ii) a twofold increase of cell volume at the end of one subculture period (two weeks), a 'Gran enano' cell suspension can give rise to between 14,580 and 100,980 plants, while between 27,000 and 117,000 plants can be regenerated from an 'Orishele' suspension.

Table: Genomic constitution, cultivar, type and number of plants obtained per millilitre of settled cells from scalp-derived embryogenic cell suspensions

Genomic constitution	Cultivar	Type	Number of plants (× 10^4) per ml settled cells
AAA	Gran enano	Cavendish	0.81-5.61
AAA	Williams JD	Cavendish	4.12-10.15
AAB	Agbagba	Plantain	0-6.02
AAB	Orishele	Plantain	1.50-6.50

Bottlenecks

Unlike dicots and seedsetting monocots, embryogenesis in bananas is very difficult. An extensive material preparation phase is required for preparing scalps for induction of somatic embryogenesis, and there are other problems and inconveniences that apply to all available methods. Low (often less than 5%) and variable embryogenic responses, combined with only 20-50% calluses resulting in highly regenerable cell suspensions, hamper the smooth establishment of *Musa* embryogenic cell cultures.

Consequently, cell suspensions that do reach the phase of mass multiplication are maintained as long as possible. This implies that the health status and quality of the cell cultures has to be checked regularly. In order to visualise whether bacteria are present, a sample of cells is transferred onto bacteriological medium and incubated at 27°C for at least 6 weeks.

Since the regeneration potential of cell suspensions decreases with time, it is recommended that embryos should be germinated regularly and plants grown subsequently from the cell cultures. Flow cytometry analysis of embryogenic cell suspensions (in collaboration with IAEA, Seibersdorf) indicates that the loss of regeneration capacity could be associated with changes in DNA content. True-to-typeness of cell suspension-derived plants varies widely, but with existing technologies field evaluation is the only reliable system.

The rate of somaclonal variation can be very low (2% in the diploid acuminata 'IRFA 903'), zero in 'Kamaramasenge' (unpublished data) or extremely high (99% in 'Williams' line E4000, unpublished data). Considering the labour-intensive and time-consuming development of *Musa* embryogenic cell cultures, and taking into account their loss of embryogenic capacity, part of the newly established cell suspensions need to be cryopreserved for backup purposes.

Applications

Mass clonal propagation

Multiple shoot clumps cultured on media with high cytokinin content have a multiplication factor ranging between 2 and 5 per month. Embryogenic cell cultures have a similar multiplication rate (4-6 times per month). However, the high regeneration capacity of embryogenic cell suspensions and the faster manipulation of liquid cultures make mass clonal propagation through somatic embryogenesis more attractive. Indeed, 1 ml settled cells of a highly regenerable cell suspension can yield up to more than 100,000 plants.

Protoplast culture

Protoplast cultures are essential for transformation through electroporation and somatic hybridisation. Like other monocot systems, embryogenic cell suspensions are the material of choice for the isolation of regenerable protoplasts in banana. Regeneration of whole banana plants derived from embryogenic cell suspensions was reported for the cv. Bluggoe (ABB group) and Grande Naine (AAA group). In all cases, a high protoplast yield is obtained after enzymatic digestion of the cell wall. Plant regeneration through somatic embryogenesis can be obtained after culture on feeder layers or using high plating densities.

Genetic engineering

Classical breeding in *Musa* has always been limited since most cultivated varieties have lost the ability to set seeds. For efficient generation of disease resistant/tolerant *Musa* spp., one often relies on genetic transformation techniques. Although very high transient expression frequencies can be obtained through protoplast electroporation (up to 1.8%), currently, genetic engineering of banana mainly proceeds with cells using particle bombardment and *Agrobacterium*-mediated transformation.

Despite the rather labour-intensive and time-consuming establishment of embryogenic cell suspensions, cell cultures are still the most suitable target material. Embryogenic cell suspensions are also suitable for gamma irradiation because of the unicellular origin of regenerated plants and thus the avoidance of chimaeras. For more information on *in vitro* mutagenesis in *Musa* embryogenic cell suspensions, the reader is referred to Roux et al.

9

Cryopreservation of Plant Tissues

The world's largest banana collection (1141 accessions) is stored at the International Musa germplasm collection at the INIBAP (International Network for the Improvement of Banana and Plantain) Transit Centre at K.U.Leuven. *In vitro* proliferating shoot tips are currently maintained under slow-growth conditions at reduced temperatures and light intensity. However, for the long-term conservation of germplasm of banana, cryopreservation of meristem cultures is considered to be the only practicable solution. Three cryopreservation protocols for meristem cultures have been developed. This paper gives an overview of pros and cons for each cryopreservation method.

The initiation of embryogenic cell suspension cultures of banana is still difficult and time-consuming, irrespective of the starting material used. Moreover, once established, these cell suspensions are subject to somaclonal variation and microbial contamination, and a prolonged culture period may result in total loss of morphogenic capacity. Up to now, most successful banana transformation procedures rely on embryogenic cell suspensions. The safe preservation of these valuable suspensions through cryopreservation is thus of utmost importance. A cryopreservation technique has been developed which involves cryoprotection followed by slow freezing and plunging into liquid nitrogen.

Currently, 51 independent cell lines and 15 different cultivars are in safe long-term liquid nitrogen storage. Recently, banana cell suspensions were recovered after 5 years storage in liquid nitrogen. Their competence for *Agrobacterium*-mediated transformation was tested. Both transient â-glucuronidase frequency and stable transformation frequency were unaffected.

INTRODUCTION

Seed preservation, the most convenient method to preserve plant germplasm, is not applicable to edible (seedless) banana cultivars. Field collections lose germplasm (genetic erosion) because of pests, diseases and adverse weather conditions, and their maintenance is labour-intensive and expensive. Therefore, *in vitro* collections have been established. The maintenance of such *in vitro* banana collections is labour intensive even under reduced growth conditions, and there is always the risk of losing accessions due to contamination or human error. Moreover, *in vitro* material is subject to somaclonal variation. Cryopreservation or freeze-preservation at ultra-low temperatures (-196°C) is the method of choice, since under these conditions biochemical and most physical processes are arrested. As such, plant material can be stored for unlimited periods. Currently, three cryopreservation protocols are available for shoot-meristematic tissues of banana: (i) simple freezing of proliferating meristem cultures using a sucrose preculture ; (ii) vitrification of apical meristems ; and (iii) vitrification of sucrose-precultured meristem cultures . Very recently it has been shown that cryopreservation could also be used for eradication of viruses . Cryopreservation is also used to conserve tissues with specific characteristics such as medicinal and alcohol-producing cell lines, genetically transformed tissues and transformation-competent tissues.

The most successful methods for improvement of banana through genetic engineering rely on the availability of embryogenic cell suspensions. Embryogenic cell suspensions are so far the only source of regenerable protoplasts in banana . When subjected to electroporation they give rise to high transient expression frequencies of an introduced marker gene . Walled suspension cells are successfully transformed by means of particle bombardment and *Agrobacterium* . Since plants regenerated from embryogenic cell suspensions are of unicellular origin they are probably also highly suitable for improvement through induced mutation.

An important drawback for the application in banana of improvement techniques based on cell suspensions remains the availability of suspensions of good quality, i.e. homogeneous embryogenic cell suspensions with high regeneration frequency. The initiation of these suspension cultures is difficult and time-consuming, irrespective of the starting material used (immature male flowers, immature zygotic embryos or proliferating *in vitro* meristems) . Consequently, it is of utmost importance to conserve embryogenic cell suspensions as stably as possible once they are established. This can be achieved through cryopreservation. Here we present a practical approach for cryopreservation of meristem cultures for long-term storage of banana germplasm. Secondly, an overview is given of the different embryogenic cell lines of banana which are currently stored in liquid nitrogen. Finally, we show that cryopreserved cell suspensions are suitable for *Agrobacterium*-mediated transformation.

MATERIALS AND METHODS

Cryopreservation of Meristem Cultures

Simple freezing of proliferating meristem cultures

Highly proliferating meristem clumps are cultured for 2 weeks on a proliferation medium enriched with 0.4 M sucrose. Surviving/ growing meristem clumps are then excised, transferred to 2 ml cryotubes and plunged into liquid nitrogen for storage. Rapid thawing takes place in a warm water bath at 40°C. Meristem clumps are recovered in a liquid regeneration medium.

Vitrification of sucrose precultured meristem cultures

Sucrose precultured meristem cultures are treated for 20 min with a loading solution (LS) which contains 2 M glycerol and

0.4 M sucrose at room temperature. Then they are subjected for 2 h to ice-cooled PVS2 solution [30% (3.26 m) glycerol, 15% (2.42 m) ethylene glycol (EG), 15% (1.9 m) DMSO and 0.4 M sucrose], transferred to 2 ml cryotubes and plunged into liquid nitrogen. Rapid thawing takes place in a warm water bath at 40°C. Directly after thawing, the toxic PVS2 solution is removed and replaced by the deloading solution (DS) for 15 min at room temperature. The deloading solution consists of 1.2 M sucrose dissolved in MS medium (pH 5.8). Meristem clumps are recovered on semi-solid regeneration medium.

Vitrification of apical meristems

One millimetre-sized meristems are excised using a binocular microscope from *in vitro* plants that had been cultured for 1 month on 6% sucrose. The most essential but also most difficult step is the excision of tiny meristem tips of the correct size (a base diameter of 0.5-1 mm). The apical domes may only be covered for two-thirds by the youngest leaf primordium. Tips that are slightly damaged or are not at the correct stage (the meristem is covered too much or too little by leaf primordia) are excluded. The tips are treated for 20 min with the loading solutions, followed by a 15 min treatment with PVS2 at 0°C. Freezing takes place in 2 ml cryovials. After storage meristem tips are rapidly rewarmed in a warm water bath (40°C) and treated for 20 min with the deloading solution. Recovery takes place on semi-solid regeneration medium. A more detailed description of the three cryopreservation methods is presented in the INIBAP Technical Guidelines 5, Cryopreservation of *Musa* germplasm .

Cryopreservation of Embryogenic Cell Suspensions and Transformation

Freezing

Suspensions are initiated from different explants: from immature male flowers , and from proliferating meristem cultures . Suspensions are cryopreserved using slow freezing (1°C/min) in the presence of

7.5% dimethylsulphoxide (DMSO) and 140 g/l sucrose. See for a more detailed description of cryoprotection, freezing, thawing and recovery of *Musa* germplasm.

Post-thaw observations and re-initiation of embryogenic cell suspensions

Two months after thawing, Petri dishes are screened for the presence of embryogenic callus. If embryogenic callus is absent, the corresponding tubes in the liquid nitrogen tank are discarded. If present, the regrowing callus is transferred to liquid maintenance medium and regularly subcultured.

Transformation

Embryogenic cell suspensions of banana were transformed according the method described by Pérez Hernandez . In this experiment, a suspension line of the plantain cultivar Three Hand Planty [THP (7+8)] was used. This suspension was established from proliferating meristem cultures in July 1994 and frozen and stored in liquid nitrogen at different time intervals (10 April, 1996; 27 September, 1996; and 22 May, 1997). On 21 December, 1999 the suspensions were thawed and plated on semisolid media. Later on, liquid cell suspension cultures were re-established from regrowing embryogenic calli. These three cell lines, together with a control (unfrozen) suspension of the same cell line, were subjected to transformation on the 9 November, 2000.

The *Agrobacterium tumefaciens* strain EHA101 contains the pFAJ3000 plasmid. This plasmid has an intron-interrupted *gusA* gene (reporter gene) under the control of a CaMV 35S promoter, and the *neo* gene coding for neomycin (selectable marker) phosphotransferase driven by the nopaline synthetase promoter (nos). Transient gene expression in banana cells is visualized after 6 days of co-cultivation with *Agrobacterium* as the number of blue spots after a 4-hour treatment with X-Gluc. After 3 months of culture on selective medium (semisolid suspension maintenance medium supplemented with 50 mg/l geneticin), independent putative transformed cell clumps are counted and transferred to regeneration medium. After 3 months, the number of regenerated plants was counted.

RESULTS

Cryopreservation of Meristem Cultures

Forty-four banana accessions obtained from the International Musa Germplasm collection at the INIBAP Transit Centre at K.U.Leuven were screened for their response towards one or more cryopreservation methods. The results are grouped according to the genomic constitution and cryopreservation method used (Table). We observed that the frequency of post-thaw regeneration depends on (a) genomic group, (b) cryopreservation protocol and (c) cultivar, despite high variations between repetitions. ABB cultivars respond very well to the simple cryopreservation protocol, whereas for AAB, AAA, some AAB plantains and AB bananas, vitrification of proliferating cultures is highly efficient. The response to vitrification of apical meristems is cultivar-independent.

Table: Response (% post-thaw regeneration) of different banana cultivars according to the genomic constitution towards different cryopreservation protocols

Group[a]	Simple freezing	Vitrification of proliferating cultures	Vitrification of apical meristems
ABB	30-68	22-58	35-86
AABpl	0-14	18-40	27-70
AAB	19-34	24-34	Nd[b]
AAA	18-31	35-68	45-80
AAAh	0-9	3-22	35-50
AB	11-35	15-34	Nd[b]
AA	15-20	10-21	Nd[b]
Ensete	0	0	38-45

[a] AABpl: Plantain; AAAh: East African Highland banana

[b] Not determined

A calculation was made of the number of accessions that can be cryopreserved per person per year according to the cryopreservation method. The time required is very dependent on the cryopreservation method. The most labour-intensive method (vitrification of apical meristems excised from rooted *in vitro* plants) would thus only be applied to cultivars which are difficult to cryopreserve with the other two methods (for example AAA highland bananas).

***Table: Comparison of three cryopreservation protocols for* Musa**

Freezing method	Labour time (h)[a]	Cvs per person per year[b]	Time needed (months)[c]	Applicable to	Expected regeneration frequency (%)
Simple Freezing	30-35	59-50	6-7	ABB group	30-70
Vitrification of proliferating clumps	40-50	49-39	12-13	AAA, AAB, some AAB plantains,...	20-70
Vitrification of individual meristems	70-80	29-25	8-9	AAAh bananas, some AAB plantains,...	20-60

[a] Calculated time needed to prepare the culture media and meristem cultures followed by cryopreservation. For each cultivar three repetitions would be made with three cryotubes per repetition containing 10 (vitrification of individual meristems) to 20 (simple freezing and vitrification of proliferating clumps) explants. For each repetition a representative sample would be thawed and regenerated.

[b] 1 year = 220 working days (8 h/day).

[c] Starting from two tubes until this accession is safely stored in liquid nitrogen (three repetitions).

Since the start of our cryopreservation work, a total of 51 accessions have now been placed 'safely' in long-term storage in liquid nitrogen using both vitrification and simple freezing.

CRYOPRESERVATION OF EMBRYOGENIC CELL SUSPENSIONS AND TRANSFORMATION

The most important parameter for post-thaw regrowth of embryogenic cell suspensions is their quality. Most suspensions which contain a high proportion of embryogenic cells (cells which are isodiametric and characterised by a relatively large nucleus, small multiple vacuoles and tiny starch and protein grains) survive cryopreservation. 'Less ideal' suspensions which contain a low proportion of embryogenic cells or in which 'embryogenic' cells are highly vacuolated or contain big starch grains seldom survive the freezing process. Since these 'less ideal' suspensions are also less favourable for transformation, not much effort is made to store them in liquid nitrogen. Currently, 51 cell lines belonging to 15 banana cultivars, representing 745 cryotubes, are safely stored in liquid nitrogen.

Table: Number of cryotubes containing embryogenic banana cell suspensions that are currently stored in liquid nitrogen

Cultivar	Cell lines	Group	1993	1995	1996	1997	1988	1999	2000	2001	Total
Musa balbisiana	1	BB	1								1
Musa acuminata	1	AA		5							5
Bluggoe	4	ABB	4				14		6		24
Monthan	1	ABB	2								2
Grande Naine	9	AAA			8	6	6		53	49	122
Gros Michel	1	AAA								16	16
Williams	13	AAA		4	25		23	8	121	48	229
Bise Egome	2	AABp					22				22
French Sombre	1	AABp		2							2
Three Hand Planty	9	AABp	1	8	34	14	16		12	7	92
Nakitengwa	2	AAAh					28				28
Orishele	4	AABp							16	112	128
Obino L'Ewai	1	AABp							8	40	48
Agbagba	1	AABp								8	8
Dominico	1	AABp					6			12	18
Total	51	15	8	19	67	20	115	8	216	294	745

AABp: Plantain; AAAh: East African Highland banana

Since transformation requires embryogenic cell suspensions in liquid media, regrowing calluses were reinitiated in liquid media. Most of the cell suspensions thus obtained displayed embryogenic characteristics comparable with the non-frozen controls. Different embryogenic cell lines of the cultivars Williams and Three Hand Planty, which had been stored in liquid nitrogen up to 4.5 years, were thawed and reinitiated. As expected, regeneration capacity was not reduced by the long storage period.

We checked whether the storage of embryogenic cell suspensions in liquid nitrogen has an effect on their transformation capacity. The cell line THP (7+8) [Three Hand Planty (AAB plantain)] was subjected to *Agrobacterium*-mediated transformation. Prolonged storage in liquid nitrogen (2.5-3.5 years) does not affect the transient *gus* expression of re-initiated embryogenic cell suspensions.

Table: Transient GUS expression of one control and three cryopreserved embryogenic cell suspension of the cv. Three Hand Planty (AAB plantain) after* Agrobacterium*-mediated transformation

Suspension line	Average no. of blue spots/ 66 µl SCV (n = 4)[a]
THP(7+8), no cryopreservation	988
THP(7+8), cryopreservation 10/04/1996, thawing 21/12/1999	1196
THP(7+8), cryopreservation 27/09/1996, thawing 21/12/1999	931
THP(7+8), cryopreservation 22/05/1997, thawing 21/12/1999	1361

[a] Numbers are not significantly different by the Tukey honest significant difference (HSD) test at an alpha level of 0.05.

Stable transformation data were also gathered. For this, the number of independent putative transformed cultures after selection on 50 mg/l geneticin, as well as the frequency of these cultures giving rise to shoots, were counted. For 2 out of 3 cryopreserved suspension cultures, the number of independent putative transformed cultures is not significantly different from the non-frozen suspensions. THP (7+8), cryo 22/05/1997, results in a significantly lower number of surviving clumps despite its comparable transient GUS expression. This might be due to the inferior physiological state of this suspension at the time of transformation. However, the calculated average of the independent putative transformed cultures of the three cryopreserved cell lines (40.7) is almost identical to that of the non-cryopreserved suspension (42.0). To test whether cryopreservation had indeed affected the stable transformation frequency, this re-initiated suspension should be tested again. More important is that all suspensions showed a comparable number of transformed shoots.

Table: Stable transformation of one control and three cryopreserved embryogenic cell suspension of the cv. Three Hand Planty (AAB plantain) after* Agrobacterium*-mediated transformation and selection on 50 mg/l geneticin

Suspension line	Average number of independent putative transformed cultures/ 66 µl SCV (n = 4)[a]	Average number of independent putative transfomed shoots/ 66 µl SCV (n = 4)	Regeneration frequency (%)
THP(7+8), no cryo	42a	20	47
THP(7+8), cryo 10/04/1996	51a	19	37
THP(7+8), cryo 27/09/1996	46a	12	25
THP(7+8), cryo 22/05/1997	25b	14	57

[a] Numbers followed by the same letter are not significantly different by the Tukey honest significant difference (HSD) test at an alpha level of 0.05

DISCUSSION

Cryopreservation of plant material serves two main purposes: conservation of genetic diversity; and conservation of material with specific characteristics. For example, large scale cryopreservation of embryogenic cultures is essential for advanced forestry breeding programs. Approximately 5000 genotypes representing 14 conifer species have been stored . For species like banana, where the establishment of embryogenic cultures is far from routine, meristem cultures are used for safe storage of germplasm. Three protocols for cryopreservation of meristem cultures are presented, which are used routinely at the INIBAP Transit Center. The method of choice mainly depends on the genome configuration of the accession. We have calculated that on average, 50 banana accessions can be cryopreserved per person per year.

Cryopreservation of banana embryogenic cultures is of utmost importance, since it is the only way for safe preservation of transformation-competent tissues. Successful cryopreservation of plant embryogenic cultures has been reported using vitrification, direct immersion in liquid nitrogen , encapsulation dehydration, encapsulation vitrification , encapsulation combined with slow freezing and vitrification combined with slow freezing . However, for non-organized tissues like cell suspensions and callus cultures, slow freezing in the presence of cryoprotective agents is still the method of choice. During slow freezing, crystallisation is initiated in the extracellular spaces, since it is believed that cells rarely contain ice nuclei. Since only a fraction of the water in the extracellular solution undergoes transition into ice, the solution becomes more and more concentrated and thus hypertonic to the cell. To restore the osmotic equilibrium, cellular water will leave the protoplast. Slow freezing will thus result in a concentrated cytoplasm which will vitrify upon subsequent plunging into liquid nitrogen. The addition of penetrating substances like DMSO facilitates this process. Vitrification (or glass formation) of the cellular solution is an important requirement for successful cryopreservation, since this is the only way to freeze a solution without the formation of lethal ice crystals.

After cryopreservation, a thawed suspension needs to be viable, able to give rise to an embryogenic cell suspension, be true-to-type, and retain its characteristics. The first two requirements are met for most embryogenic cell lines of banana. The true-to-typeness of plants recovered from cryopreserved embryogenic cell suspensions of banana was screened by Côte and co-workers . They proved that there was no difference at the agronomical level between plants produced from cryopreserved embryogenic cell suspensions and control plants. Finally, a cryopreserved suspension needs to retain typical features. It was proven that cryopreservation did not affect the expression of a foreign *sam* gene in transgenic *Papaver somniferum* cells , had no impact on paclitaxel biosynthesis in suspension cultures of *Taxus chinensis* , and was even beneficial for pyrethrin biosynthesis in *Chrysanthemum cinerariaefolium* Vis. cells . Also, the production of regenerable protoplasts was not influenced by cryopreservation as shown for rice, and *Festuca* and *Lolium* species . Moreover, cryopreserved rice callus and maize cell cultures proved to be a constant source of regenerable cell cultures for the production of transgenic plants. Moukadiri and co-workers showed that calluses that were stored in liquid nitrogen demonstrated a higher competence for transformation, as indicated by transient gene expression levels. In banana, comparable levels of transient expression, as well as stable transformation, were obtained in both cryopreserved and non-cryopreserved suspensions. We can therefore that assume that there was no cryoselection.

10

MUTATION INDUCTION IN MUSA

Induced mutation techniques are particularly important for bananas and plantains (*Musa* spp.) where there is limited sexual reproduction that could generate genetic variation, the basis for selection. Although spontaneous mutations have contributed to the genetic diversity of *Musa* and significantly increased its variation, their frequency is very low. The use of *in vitro* cultures for induction of mutations in *Musa* spp. could be the method of choice if several steps of the mutation induction process could be optimised. The following aspects were investigated: (a) the plant material used; (b) the determination of an optimal mutagenic dose for physical or chemical mutagens; (c) the dissociation of chimerism; and (d) the application of an early mass screening method for the selection of useful mutants with improved characters. The proposed new methodology, which will speed up the mutation-induction process and improve its efficiency, may revolutionize the use of mutations in *Musa* spp., especially at a time when there is an increased interest in collecting mutants to understand gene structure and function.

INTRODUCTION

Genetic variation is the starting point of any breeding programme. Genetic variation may already be present in nature, may be obtained after several years of selection, or may be produced through hybridisation (for seed propagated crops). For vegetatively propagated crops such as edible *Musa* spp. it is difficult to obtain genetic variation, because of sterility and polyploidy. Spontaneous somatic mutations have played an essential role in the speciation and domestication of plantains and bananas.

Some authors even claim that all the bananas and plantains that we grow and eat were selected in prehistory from spontaneous mutations . Some of these spontaneous mutants in *Musa* (members of the Cavendish subgroup and plantains) now play a very important role in banana and plantain production.

Unfortunately, the rate of occurrence of spontaneous mutations is too low to satisfy practical breeding needs. Mutagenic agents such as radiation and certain chemicals can be used to induce mutations at a higher frequency and generate genetic variation from which desired mutants may be selected. Stover and Buddenhagen , Van Harten and De Langhe proposed mutation techniques as a tool for plantain and banana improvement.

MATERIAL

Mutation Induction using *in Vivo* Plant Material

Early studies on mutation induction in plantains and bananas were performed on seeds and suckers. Stotzky et al. reported on the effects of gamma rays on seed germination and seedling survival in the diploid *Musa balbisiana.* Reduction of sprouting by irradiation was observed with 'Gros Michel' suckers and there was no survival with doses higher than 50 Gy.

One mutant with leaf aberrations and another with more intensive pigmentation were selected in the M_1V_2 generation. Several researchers suggested the use of mutagenic agents for induction of resistance to several diseases.

They also suggested using adventitious buds for regenerating banana. Menendez applied ethyl methanesulphonate (EMS) to seeds of diploid *Musa acuminata* in Jamaica . Seeds were soaked for 48 h in 0.04-0.20 M EMS at a temperature of 20°C. A concentration of 0.10-0.14 M reduced germination by 50%.

Irradiation of suckers only gave a low yield of mutagenized material for further screening. Banana suckers have been irradiated with gamma rays prior to excision of shoot tips and *in vitro* culture. A dose of 10 Gy stimulated explant growth and shoot formation. Leaf expansion on regenerated plantlets was observed up to 25 Gy, but 50 Gy was inhibitory.

However, De Guzman et al. reported that viable shoots could be obtained from suckers irradiated with a dose up to 100 Gy. Sucker irradiation *in situ* before tissue culture initiation was not effective and yielded only a low number of regenerants in the M_1V_1 generation due to the small number of treated suckers. Even when *in vivo* suckers were irradiated before meristem tip isolation and culture, it was not very effective.

The combination of mutation breeding and *in vitro* culture (also called *in vitro* mutagenesis) has been found to make the induction and the selection of induced somatic mutations more effective.

This method offers several advantages: (a) plant material can be propagated rapidly before treatment to obtain a large population size; (b) further propagation will ensure the formation of periclinal chimeras or homohistont individuals; and (c) increased recovery through decreased somatic competition can be obtained by modifying culture conditions.

Plant growth regulators, and in particular a cytokinin-enriched medium, can increase the recovery rate of mutated cells. Hence the combined use of mutation induction and *in vitro* technology is more efficient, because it speeds up the production of mutants as a result of an increased propagation rate and a greater number of generations per unit time and space.

METHODOLOGY

The *Musa* mutation-induction system, based on *in vitro* techniques to recover mutant plants and micropropagate desirable mutants, was developed by Novák et al. in the IAEA Laboratories. It is now applied worldwide in several *Musa* breeding programmes.

MUTAGENIC TREATMENT

Physical Mutagenesis

Gamma irradiation is the main physical mutagen used to induce genetic variation. Novák et al. , Espino et al. and Epp described the dose response of tissue-cultured shoot tips to gamma irradiation. More recently Roux standardised the methodology to provide guidelines to mutation induction programmes in *Musa* spp.

Shoot tips excised from seven clones representing the different genomic constitutions of the genus *Musa* [(Calcutta-4 (AA), Kamaramasenge (AB), Tani (BB), Grande Naine (AAA), Williams (AAA), Three Hand Planty (AAB) and Cachaco (ABB)] were treated with 10 doses from 10 to 100 Gy of a ^{60}Co gamma irradiation source at a dose rate of 44 Gy/min.

For each *Musa* accession, 200 explants were treated for sensitivity testing and 20 non-irradiated explants were used as controls. Immediately after irradiation, the explants were placed onto fresh semi-solid medium (MS with 20 μM BAP). Radiation sensitivity and post-irradiation recovery were assessed by measuring the following parameters:

- The survival rate: measured 40 days after irradiation; a non-surviving explant is an explant which is completely brown and has no shoots.
- The propagation rate: measured 40 days after irradiation. All shoots per surviving explant were counted.
- The shoot height: measured 40 days after irradiation. Only the highest shoot of the surviving cluster is counted, from the base of the explant up to the intersection of the youngest leaf with the next leaf. The height is measured in ranges of approximately 0.5 cm.
- The fresh weight: measured every 10 days until 40 days after irradiation. Five explants were placed in a sterile Petri dish and weighed under sterile conditions.

All parameters were expressed as percentages of the control.

The current *in vitro Musa* mutation-induction process using shoot-tip culture and field selection (modified from).

Steps	Time
Sucker from the field	T_0
Shoot tip culture ⇓	T_0
Radiosensitivity test (minimum 200 shoot-tips) ⇓	T_0 + 6 months
Mutagenic treatment of minimum 2000 shoot tips with a LD_{50} Dose ⇓	T_0 + 12 months
Micropropagation M_1V_1 ⇓	T_0 + 13 months
Micropropagation M_1V_2 ⇓	T_0 + 14 months
Micropropagation M_1V_3 ⇓	T_0 + 15 months
Rooting M_1V_4 ⇓	T_0 + 16 months
Acclimatization to soil ⇓	T_0 + 17 months
Field Selection ⇓	T_0 + 24 months
Genetic confirmation and agronomic evaluation ⇓	T_0 + 48 months
Micropropagation of desired plants ⇓	T_0 + 60 months
Multilocation trials	T_0 + 84 months

LD_{50}: Lethal dose for 50% survival
M_x: Mutagenic treatment
V_x: Vegetative generation
T_0: Time when the process is initiated

The different *Musa* accessions showed different responses according to their ploidy level and genomic constitution. Diploid AA and BB clones were more sensitive than triploid clones. In general the A genome was more sensitive than the B genome. The cultivar Three Hand Planty (AAB) showed the highest sensitivity among the triploid clones, but this could be due to its severe blackening caused by oxidation, which makes it difficult to measure survival rates. Among the seven clones studied, the clone Cachaco (ABB) expressed the lowest sensitivity to gamma irradiation. The two AAA genomes, Williams and Grande Naine, had a similar response to gamma radiation. The clone Kamaramasenge (AB) behaved more like a triploid clone than a diploid clone. Flow cytometric analysis revealed that this Kamaramasenge clone was indeed a triploid. The optimal dosage for mutation induction depended on the parameter studied. For example the survival rate always indicated higher optimal doses than the multiplication rate. Among all the parameters studied, survival rate and fresh weight are preferred because they result in less experimental error than shoot height and multiplication rate. Nevertheless, explant survival is not always clear and fresh weight measurements are labour-intensive and can cause contamination.

The lethal dose at 50% (LD_{50}) can thus no longer be used as a reference. The tendency is now to use relatively low doses because they produce less chromosomal damage and other negative side effects than stronger treatments. However this results in lower mutation frequencies.

The following ranges of doses are recommended:

- 10-20 Gy of gamma irradiation for diploid clones Calcutta4 (AA) and Tani (BB).
- 30-40 Gy of gamma irradiation for the triploids Three Hand Planty (AAB), Grande Naine (AAA), Williams (AAA) and Kamaramasenge which is a triploid.
- 40-50 Gy of gamma irradiation for the triploid Cachaco (ABB).

Chemical Mutagenesis

Shoot tips propagated *in vitro* have also been used for chemical mutagen treatments. Omar et al. studied the effects of ethyl methanesulphonate (EMS) on growth and development of excised shoot tips of two banana clones: SH-3362 (AA); and GN-60A, a mutant of Grande Naine (AAA). The effects of EMS on fresh weight and on the number of newly initiated adventitious buds of cultured shoot tips were clear after 30 days of incubation. A suitable level of EMS treatment was 24.69 mM for 3 h for both banana clones . Epp , on the other hand, suggested incubation of shoot-tips for 1 hour in 12.41-37.23 mM EMS. Bhagwat and Duncan compared the effect of three chemical mutagens, namely sodium azide (NaN_3), diethyl sulphate (DES) and EMS, at various concentrations on shoot tips of *in vitro* grown cultures of banana (*Musa* spp., AAA Group cv. Highgate). They considered two criteria: the number of apices that survive the treatment and the number of regenerated shoots, coupled with the factor of effectiveness (FE) using the following formula :

$$FE(\%) = \frac{\text{Total number of variations}}{\text{Total number of apices treated}} \times 100$$

For NaN_3, the highest FE (7.8%) was obtained by the 30 min/ 2.3 mM treatment, which resulted in 63.3% explant survival and 58.9% shoot regeneration. For DES, a FE of 7.9% was obtained with the 60 min/20 mM treatment, which resulted in 65.5% survival and 38.2% shoot regeneration from explants. For EMS mutagenesis, the recommended treatment is 30 min/200 mM, which resulted in a FE of 5.8%, 80% explant survival, and shoot regeneration from 31.6% of apices. This recommended dose and time of treatment for EMS is quite different from those mentioned above. However, a longer incubation period coupled with low concentrations is preferred because it decreases damage causing hydrolytic by-products and thus improves the mutagenic efficiency .

MONITORING CHIMERA DISSOCIATION AFTER MUTAGENIC TREATMENT

In vitro mutagenesis of multicellular meristems of *Musa* spp. leads to a high degree of chimerism. Repeated vegetative propagation must be carried out to dissociate chimeras, but the minimum number of cycles required is unknown. In general, mutated cells are difficult to monitor, but mutations, which result in a change in genome number, are an exception. Roux et al. simulated this case by colchicine treatment, followed by flow cytometric analysis. Colchicine treatment induced ploidy chimerism (mixoploidy), and chimera dissociation was assessed using three different propagation systems (shoot-tip culture technique, multi-apexing culture technique and corm slice culture technique). The average percentage of cytochimeras was reduced from 100% to 36% after three subcultures using shoot-tip culture, from 100% to 24% when propagating by the corm slice culture technique, and from 100% to 8% after the same number of subcultures using the multi-apexing technique. None of the propagation systems eliminated chimerism completely.

CHARACTERS IMPROVED

From the FAO/IAEA mutant varieties database, two banana accessions were registered as improved mutant varieties: 'Novaria' and 'Klue Hom Thong KU1'.

Novaria

Using the shoot-tip mutation-induction system (Figure 1), an early flowering putative mutant designated 'GN-60A' was identified among the regenerated plants obtained from gamma ray-treated populations in the glasshouses of the FAO/IAEA Agriculture and Biotechnology Laboratory, Seibersdorf, Austria. The GN-60A clone has been micropropagated and its vegetative progeny sent to Honduras, Australia, South Africa and Malaysia for field testing under commercial plantation conditions. Reports from these four sites showed that not all the plants demonstrated earliness . The chimeric constitution of the original plant could explain this behaviour, as only progenies deriving from the mutated sector are supposed to give rise to the early putative mutant. In Malaysia, early flowering plants obtained in the field were tissue cultured again to produce about 2000 plants for commercial evaluation in the United Plantations Bhd. in September 1993. From this population a selection, flowering about 10 weeks earlier than the original parental clone (i.e. Grande Naine), was launched under the name Novaria in 1995.

Klue Hom Thong KU1

In Thailand, at the Department of Horticulture, Kasetsart University in Bangkok, a mutant variety Klue Hom Thong KU1 was released . This variety was obtained by treating tissue cultures with gamma rays at 2.5 krad (25 Gy). The mutant was selected for its bunch size and cylindrical shape from which larger bananas can be selected.

Other desirable variants and putative mutants have been identified for release or further confirmation trials. As shown in Table below, most of the selected traits are agronomically improved characters.

Table: List of putative mutants obtained

Country	Parent clone	Selected clone	Selected traits	Technique	Place of induction
Cuba	SH-3436	SH-3436-L9	Height reduction	Gamma rays	Cuba
	Paredido al Rey	6.44	Height reduction	Gamma rays	IAEA
Malaysia	Pisang Rastali	Mutiara	FOC tolerance[a]	Somaclonal variation	Malaysia
	Novaria		FOC tolerance[a]	Somaclonal variation	Malaysia
Philippines	Lakatan	LK-40	Height reduction	Gamma rays	IAEA
	Latundan	LT-3	Larger fruit size	Gamma rays	IAEA
Sri Lanka	Embul	Embul-35 Gy	Earliness	Gamma rays	Sri Lanka
IAEA	Grande Naine	GN35-I to GN35-VIII	Tolerant to toxin from *Mycosphaerella fijiensis*	Gamma rays	IAEA

[a] FOC, *Fusarium oxysporum* f.sp. *cubense*

Disease resistance seems to be more difficult to obtain through mutation induction techniques. Smith et al. used an original strategy to obtain disease resistance. Instead of irradiating an agronomically superior genotype that is disease susceptible, they irradiated Dwarf Parfitt, an extra dwarf Cavendish banana that has shown a high level of resistance to race 4 of Fusarium wilt. Following radiation, 35 M_1V_3 out of 500 explants irradiated at 20 Gy were recovered that possessed improved agronomic characteristics (taller plant size, increased yield and no choking). Most importantly these selections appeared to retain the resistance to race 4 shown by the mother plant Dwarf Parfitt.

Bhagwat and Duncan [21,29] used gamma irradiation (8-20 Gy) and chemical mutagens to obtain Highgate (AAA) tolerance to *Fusarium oxysporum* f.sp. *cubense.* Twelve weeks after inoculation in the greenhouse, 0.3-0.9% of the regenerated plants from irradiated explants and 1.9-6.1% of chemically treated explants had less than 10% vascular invasion in their corms with no external symptoms. These plants were considered tolerant to the fungus and were multiplied, *ex vitro*, for field screening.

FUTURE PROSPECTS AND CONCLUSIONS

Even though the traditional shoot-tip mutation-induction techniques applied to *Musa* spp. improvement has produced some useful mutants, there are some limitations, i.e. treated shoot tips show a high degree of chimerism. This is a serious obstacle because mutated cells cannot be distinguished from non-mutated cells. Hence techniques that reduce the formation of non-chimeric plants will enhance the production of more stable and useful mutants. The advancement of mutation induction relies greatly on embryogenic cell suspensions .

Another obstacle comes from the random nature of mutation induction. Hence as many plants as possible need to be screened, but in bananas field screening is expensive, laborious and site-specific. Consequently selection efficiency would be increased by the development of mass screening methods at the *in vitro* level followed by further testing of selected plants in the greenhouse and in the field.

In recent years interest increased substantially to understand the *Musa* genome. This is caused by new advances in research and the access to enabling technologies to understand gene structure and function. Mutation induction can obviously contribute to functional genomics.

11

USEFULNESS OF EMBRYOGENIC CELL SUSPENSION CULTURES

In bananas shoot-tip cultures are traditionally used for mutagenesis. The main problem with this type of culture is the presence of chimerism. In contrast, embryogenic cell suspensions allow the handling of large populations under controlled conditions and simultaneously avoid chimerism if embryos have a single-cell origin. In this study, we verified the unicellular origin of somatic embryos by treating embryogenic cell suspensions (ECS) with colchicine and determining the ploidy of regenerated plants by flow cytometry. We have found that none of the plants regenerated from colchicine-treated ECS was mixoploid (chimeric). To optimise a protocol for the irradiation of ECS, several experiments were performed. Flow cytometric analysis of cell-cycle kinetics proved to be more informative than growth curves for the determination of the optimal timing of irradiation. The optimal irradiation dose was determined using three different parameters. Among them, the fresh weight gain and the regeneration capacity indicated that the optimal irradiation dose lies between 50 and 75 Gy for embryogenic cell suspensions of both Williams (AAA) and Three Hand Planty (AAB plantain).

INTRODUCTION

Bananas and plantains (*Musa* spp.) are two of the world's major crops. They are a staple food of millions of people and rank among the top five food commodities. Only 13% of the total banana production, mainly dessert bananas, is produced for export . Hence, bananas and plantains are important components of food security in the tropics. The three main diseases and pests that significantly affect banana cultivation are Fusarium wilt, black Sigatoka and nematodes. These diseases/pests can reduce fruit yield by 50%.

No fungicides can be applied against Fusarium wilt. For black Sigatoka and nematodes, chemical control strategies exist but are environmentally unsound, hazardous and too expensive for many farmers. Breeding of resistant cultivars is the only sustainable way to reduce the use of pesticides. Just as evolution is based on genetic variation, so is breeding. Variation is induced by hybridisation and mutation, among other mechanisms. Breeding is difficult in edible *Musa* spp. because they are polyploid and sterile.

Almost all the edible banana and plantain varieties originate through spontaneous mutations. The best example is the spontaneous banana mutant 'Cavendish' originating from Vietnam, which is resistant to Fusarium wilt (race 1) and which replaced 'Gros Michel' in the 1950s and '60s. The discovery of this banana mutant saved the banana industry from collapsing. Consequently, the export trade relies on a very narrow genetic base since only one or two triploid *acuminata* varieties of the subgroup Cavendish dominate the export market . The banana is probably the best example in the history of agriculture of the pathological danger of monoclone culture. Indeed, without clonal diversification, the trade can hardly be expected to survive indefinitely, and the generation and use of genetic variability may be the only remaining option.

Spontaneous mutations occur at a relatively low frequency. This frequency can be enhanced using physical or chemical mutagenesis. Combination of *in vitro* mutation techniques (*in vitro* mutagenesis) raised hopes in the 1980s and '90s. However, not many useful mutants have been obtained so far. This is mainly due to the high degree of chimerism that occurs during mutagenesis of *Musa* meristems. Depending on the micropropagation method, chimerism can, however be reduced to 8% after four subcultures *in vitro*. Nevertheless, no propagation system dependent on meristematic tissues eliminated chimerism completely . In several crops, somatic embryogenesis has been shown to be an effective method for eliminating chimeras, since embryos are presumed to regenerate from single cells . This study was initiated with the aim of optimising the *in vitro* mutagenesis technique to induce genetic variation in *Musa*, using embryogenic cell suspensions as target material for irradiation. The work was divided into three phases: (a) demonstration of the single cell origin of somatic embryos; (b) determination of the optimal timing for mutagenic treatment; and (c) determination of the optimal dose of gamma irradiation.

MAINTENANCE AND REGENERATION OF EMBRYOGENIC CELL SUSPENSIONS (ECS)

The establishment of embryogenic cell suspensions (ECS) using the scalp method consists of three steps: (a) the preparation of highly proliferating material; (b) the induction of embryogenic cell aggregates; and (c) the maintenance of embryogenic cell aggregates in liquid medium. These initial steps were made at the Laboratory of Tropical Crop Improvement, K.U.Leuven, Belgium

Maintenance of ECS

The age of ECS used ranged from a few months up to 9 years. During this period, they were maintained at the Laboratory of Tropical Crop Improvement, K.U.Leuven. Then they were hand-carried to the Plant Breeding Unit of the FAO/IAEA Agriculture and Biotechnology Laboratory. Different cell lines of the following cultivars were used:

- Williams (AAA)
- Grande Naine (AAA)
- Three Hand Planty (AAB plantain)
- Bluggoe (ABB)

Individual embryogenic cultures originating from different scalps (cell lines) were kept separately. Erlenmeyer flasks were placed on an orbital shaker at 70 r.p.m. (Figure 1). The medium was refreshed every 2 weeks. The ECS were multiplied in ZZ medium (containing zeatin and 2,4 D) until a sufficient number of flasks had been obtained for further experiments.

Regeneration of ECS

Regeneration of ECS into pro-embryos, embryos and finally plantlets was done in three steps, each involving a different medium: R1 (hormone-free MS), R2 (MS containing 1 mM BAP) and R3 (MS containing 1 mM BAP and 1 mM IAA) respectively . Regeneration took place on semi-solid medium in Petri dishes for R1 and R2 media and in test tubes for R3 medium. Embryos developed in R2 were also regenerated using the Temporary Immersion System (TIS) vessels containing R3 liquid medium.

THE ORIGIN OF ECS

Biotechnological techniques such as *in vitro* mutagenesis and genetic transformation depend on the availability of efficient methods for chimera dissociation . Considering this, somatic embryogenesis is the most promising method since somatic embryos have been assumed to be of single-cell origin . In some species, histological studies confirmed the single-cell origin of somatic embryos. They develop either directly from an explant or as secondary embryos at the surface of older somatic embryos . Grapin et al. deduced from cytological studies on somatic embryo ontogenesis in *Musa* that a unicellular origin was more than likely. However, such studies can only be performed on a few somatic embryos, and thus the extrapolation of these findings to a large number of somatic embryos is risky. Mixoploidy induction with colchicine and ploidy analysis with flow cytometry allowed Roux et al. to monitor the effectiveness of three different micropropagation techniques in dissociating chimeras. This approach was applied in this study to verify the unicellular origin of somatic embryos from ECS.

Effect of Colchicine on ECS

Two cell lines of the variety Williams (AAA Group) (WIL-124C and WIL-124T) were sieved using a 1000-μm pore-size mesh to obtain a fine suspension. The ECS were then subcultured in the maintenance medium ZZ in 100 ml flasks at an initial concentration of around 3% settled cell volume (SCV). The cells were treated with colchicine for 3 days after subculture at a final concentration of 0.05%, 0.1% and 0.2% (w/v) for 24 h. The cells were then washed three times in ZZ medium. After 24 h colchicine treatment, 0.5 ml of ECS at 10% SCV was taken from each flask and transferred to a Petri dish containing the regeneration medium (R1). Nuclear DNA content distribution in the ECS of the two cell lines WIL-124C and WIL-124T was determined using flow cytometry as described previously before transfer to the regeneration medium R1 and after complete regeneration into plantlets.

Table: Ploidy distribution of regenerated plants from colchicine-treated ECS of two triploid cell lines of the cultivar Williams

Cell line	Colchicine concentration (%)	Regenerated plants	3x	6x	Mixoploidy (3x + 6x)
WIL-124C	0	5	5	0	0
	0.05	12	12	0	0
	0.1	5	3	2	0
	0.2	0	0	0	0
Sub-total		22	20	2	0
WIL-124T	0	108	108		0
	0.05	63	58	5	0
	0.1	37	36	1	0
	0.2	88	84	4	0
Sub-total		296	286	10	0
Total		318	306	12	0

Even though the two cell lines (Wil-124C and Wil-124T) were initiated from the same accession [Williams (ITC-0365)], the effect of colchicine on polyploidy induction was different. The response to colchicine treatment was clearer for cell line WIL-124C than for WIL-124T, especially when high concentrations were applied (0.1% and 0.2%). The proportion of polyploidised cells increased to about 50% for WIL-124C, but only to 15% for WIL-124T.

Regenerated Plants from Colchicine-treated ECS

The embryos were subsequently transferred to R3 medium in test tubes. As soon as green plantlets with shoot and roots were obtained, 0.5 cm^2 pieces of leaves were excised and their ploidy measured by flow cytometry before acclimatization. For both cell lines, WIL-124C and WIL-124T, the majority of the plants regenerated were triploid. ECS line WIL-124C, which had a higher proportion of hexaploid cells at the earlier stage, had a lower regeneration rate. No plants were regenerated after treatment with 0.2% colchicine. Control cells of WIL-124C, however, had also a very poor regeneration capacity. This could be due to the different quality of cells between the two cell lines.

WIL-124T had a better regeneration capacity. Nevertheless, among the treated cells, the proportion of regenerated hexaploid plants (5.3%) still remained very low compared with triploid plants. Hence we speculate that triploid cells have an advantage over hexaploid cells during culture. Interestingly there were no mixoploids among the regenerated plants, which is consistent with a single cell origin of embryos. Thus embryogenic cell suspensions seem to be the material of choice for mutagenic treatments.

IRRADIATION OF ECS

Determination of the Optimal Timing for Irradiation

Growth curve of ECS

The time at which the cells are irradiated is very critical. The growth of suspension cultures is usually divided into three phases: the lag phase, the exponential phase, and the stationary phase. Irradiation of the cells in the exponential (active growing) phase is usually recommended, to increase the chance of recovering higher numbers of mutants. To determine the exponential phase, we measured the percentage of settled cell volume (% SCV) by regularly transferring the content of Erlenmeyer flasks to sterile centrifuge graduated tubes and using the following formula:

$$\%SCV = \frac{\text{Volume of settled cells}}{\text{Total volume (cell + medium)}} \times 100$$

Five 100-ml Erlenmeyer flasks per cell line were inoculated with 15 ml of cell suspension (at an SCV of 3.3%). Growth curves were obtained by measuring the SCV twice per week (every 2-3 days) until day 29 (Figure 3). Cells were not subcultured until the exponential phase was reached, after 15 days or more. None of the five cell lines followed a sigmoid growth curve. Growth curves of cell lines of Three Hand Planty (THP1 and THP7) seemed to reach an exponential phase after 19 days. These two THP cell lines had a tendency to form clumps after some days after subculture, which could explain why their growth curves were different from the other three cell lines. The growth curves of the three other cell lines, Grande Naine (GN), Williams (WIL) and Bluggoe (BG), were linear rather than sigmoid. Additionally, cells in culture for periods longer than 2 weeks show a decreased quality. Thus, SCV did not seem suitable for the determination of the optimal timing for irradiation.

Cell-cycle analysis of ECS

Cell growth and division is a cyclical process. With flow cytometry, it is possible to study the proportion of cells in the G_1 and G_2 phases of the cell cycle, since the nuclear DNA content in G_2 is twice that in the G_1 phase. Cycling cells transit through the G_2 phase, whereas non-cycling cells usually remain in G_1 ($G_{0/1}$). Because the effect of a specific mutagen on chromosomes depends on the position of a cell in the cell cycle at the time of exposure, it is important to study the cell cycle of embryogenic cell suspensions. We therefore analysed the proportion of G_2 and G_1 cells. Cells from four accessions (four 100-ml Erlenmeyer flasks per accession) were analysed every 2 or 3 days during 30 days of culture. Peaks indicated waves of mitotic activity, i.e. a certain degree of cell-cycle synchrony. A similar mitotic activity was described by Panis by measuring the mitochondrial activity with the 2,3,5-triphenyl tetrazolium chloride reduction test during the culture of embryogenic banana cell suspensions. It seems that in cells transferred to a fresh medium, the cell cycle is stimulated relatively synchronously. The highest proportion of cells in G_2 phase was observed 8 days after subculture for Williams and the two THP accessions, and 6 days after subculture for Bluggoe. These are, however, preliminary results, and a more detailed study using synchronized cultures should be performed to confirm these findings.

Chromosome-type aberrations (involving both chromatids) are produced by ionising radiation in the G_1 phase of the cell cycle. During the S phase, the lesions are transmitted to both chromatids and are observed during metaphase or anaphase of the succeeding mitosis . According to Figure 4, the majority of cells are in the G_1 phase during the first 3 days after subculture. Interestingly, the efficiency of biolistic transformation of ECS is highest during the first 3-4 days of culture (Swennen, personal communication). We thus consider 3 days after subculture the optimal moment for mutagenic treatment.

Determination of the Optimal Mutagenic Dose

Irradiation of ECS

Two cell lines of the varieties Williams (AAA) and Three Hand Planty (AAB) were sieved using a 1000-μm pore size mesh to obtain a fine suspension. The ECS were then subcultured in the maintenance medium (ZZ) in 100-ml flasks at a concentration of around 3% of settled cell volume (SCV). Three days after subculture, 0.5 ml of cells were transferred to a sterile Petri dish and the medium was removed. The cell aggregates were then irradiated at doses ranging from 0 to 250 Gy in steps of 25 Gy using a ^{60}Co gamma source at a dose rate of 30 Gy/min. After irradiation, the cells were resuspended in fresh medium in centrifuge tubes and transferred to 100 ml Erlenmeyer flasks at different quantities according the parameter to be analysed. To study the effect of gamma radiation on the growth of ECS, settled cell volume, fresh weight gain and regeneration capacity were determined. The results were expressed as percentages of the control (non-irradiated cells) at all doses.

Radiosensitivity of ECS based on the percentage of settled cell volume

After several preliminary experiments, we observed that the variation of SCV between flasks was too high. Therefore we divided each Erlenmeyer flask into two identical subsamples each of 3% SCV. For each dose (from 0 to 250 Gy), one subsample was irradiated and the other served as control (non-irradiated). The SCV was calculated every week according to the formula given above. Clearest results were obtained after 21 days. Based on SCV, the radiosensitivity curves were quite similar for Three Hand Planty and Williams. This would indicate that the response was not genotype-dependent. Surprisingly, Figure 5 shows that even at a relatively high dose of 250 Gy (SCV_{250} = 7.8%), cells have a settled cell volume corresponding to nearly 50% of the control (SCV_0 = 14.3%). At doses over 100 Gy cells looked brownish but were still able to divide.

Radiosensitivity of ECS based on fresh weight gain

To study the fresh weight gain after irradiation, the SCV was adjusted to 10% and 0.5 ml of the homogenized suspension was distributed over a moistened filter paper saturated with the regeneration medium R_1. Four Petri dishes were inoculated per dose. For each Petri dish, the saturated filter paper was weighed before and after spreading the cells. The fresh weight of the cells was calculated using the following formula:

Fresh Weight of Cells = total weight (filter paper + cells) - filter paper weight.

Each week, during a 56-day period, the cell fresh weight was compared to the control. The optimal time to observe the effect of gamma radiation on fresh weight was 28 days after irradiation. The radiosensitivity curves for the varieties Williams and Three Hand Planty are quite similar. After 28 days, the cell-weight gain at 75 Gy (CWG_{75} = 0.84 g) was 50% of the control (CWG_0 = 1.68 g). The effect of gamma radiation on fresh weight gain thus seems to be greater than on the SCV. This difference could also be due to the different medium constitution (semi-solid versus liquid medium).

Radiosensitivity of ECS based on regeneration capacity

Approximately 40 days after irradiation and transfer onto semi-solid regeneration medium R_1, the pro-embryos with filter paper were transferred to the second semi-solid regeneration medium R_2 for about 15-20 days until embryo formation. Because of the high number of embryos, the experiment was limited to one Petri dish per dose.

The content of a Petri dish was transferred to a TIS vessel containing the third regeneration medium R3 (liquid), except for the non-irradiated cells where the content of a Petri dish was transferred to two TIS vessels because of the higher number of embryos. Later, green plantlets were counted and transferred to Magenta GA7 boxes containing semi-solid R_3 regeneration medium for further growth before acclimatization. The radiosensitivity curve was obtained by comparing the number of regenerated plantlets for each dose with the control plants (from non-irradiated ECS).

Radiation at a low level seems to stimulate regeneration, especially in Williams. We must, however, take into account that in control plants, the density of embryos in the TIS vessels was too high and hence a considerable number of small plantlets could not develop. In both genotypes no plants regenerated above 200 Gy. The number of regenerated plants drops drastically above 50 Gy for the variety Williams, whereas for Three Hand Planty the number of regenerated plantlets decreases less dramatically. The regenerated plants have recently been transferred to the greenhouse and the field for further phenotypic and agronomical evaluation.

CONCLUSIONS

Our observations clearly demonstrate that a combination of colchicine treatment and DNA flow cytometry is useful for the verification of the unicellular origin of somatic embryos. The results demonstrated that embryogenic cell suspensions (ECS) are probably the material of choice for *in vitro* mutation induction, as their use overcomes the problem of chimerism. This will speed up the *in vitro* mutagenesis process since there is no need to dissociate chimeras after mutagenic treatment. DNA flow cytometric analysis is also very powerful to study cell-cycle kinetics of ECS and to determine the optimal timing for irradiation. Three days after subculture seems to be the optimal moment for mutagenic treatment. Nevertheless, additional studies should be carried out, preferably with synchronized cells, to obtain more precise results. To study the effect of gamma irradiation, regeneration capacity seemed to be more appropriate than settled cell volume or fresh weight gain. The optimal dose for the irradiation of suspensions was determined as 50-75 Gy for Williams (AAA) and Three Hand Planty (AAB). Hence embryogenic cell suspensions from *Musa* can tolerate higher doses than shoot tips.

12

Analysis of Induced Mutants

Mutants were induced in the two most popular Philippine banana cultivars by gamma and fast neutron irradiation of *in vitro* shoot-tip cultures. Most of the clones produced were morphologically similar to the non-irradiated cultivars. Promising clones were selected and evaluated further using molecular markers. The RAPD, SSR and AFLP techniques were successfully established under local conditions and found useful for characterizing banana and other *Musa* species. RAPD and AFLP markers showed sufficient polymorphism for genotype discrimination, but SSR and AFLP markers are more highly reproducible. Only AFLP markers possess both high reproducibility and discriminatory capability. AFLP can distinguish between the two cultivars used with a minimum of one selective primer pair.

It can also detect variation in DNA profiles of induced mutant clones which are otherwise morphologically indistinguishable, and detect variation between the induced mutant parent clones and their derived suckers. A non-radioactive silver staining technique for AFLP markers suitable for local laboratory conditions has been established. The results obtained from both morphological and molecular analyses show that more point mutations were generated by irradiation with fast neutrons than with gamma rays. However, the number of vegetative propagation cycles for the shoot-tip technique used may not be sufficient to eliminate chimeras completely in the mutated populations. The results obtained could provide a sound basis for the successful application of mutation and molecular marker techniques to improve bananas in the Philippines.

INTRODUCTION

The banana is one of the most important fruit crops of the Philippines. It is important both as a source of local and international revenues for farmers, and an important component of the daily diet of all Filipinos. Two *Musa acuminata* cultivars, 'Lakatan' (AAA) and 'Latundan' (AAB), are among the most popular dessert type of bananas produced primarily by small-scale banana farmers and consumed by the local market. Their production and trade potentials are severely constrained by low yield, susceptibility to diseases and undesirable post-harvest qualities. Breeding strategies to improve these traits are difficult to implement because of the asexual behaviour, sterility and polyploidy of edible bananas .

Induced mutation (either by chemicals or irradiation), coupled with *in vitro* propagation technique such as shoot-tip culture, has been established as a tool to generate variation in a number of vegetatively propagated crops. Induced mutants are usually differentiated from each other and from the original cultivar by phenotypic analysis based on agromorphological and physiological traits. This approach could be severely limited by the large size of mutagenized populations, particularly in the case of banana, and by dominance and developmental and environmental effects in the traits used in characterizing the population. Chimeras obtained after irradiation are usually difficult to detect. However, advances in molecular marker analysis may overcome these limitations.

In banana, several DNA marker techniques have been used to investigate genetic relationships between *Musa* accessions, and to determine differences in somaclonal variants and radiation-induced mutants. These are restriction fragment length polymorphism (RFLP), random amplified polymorphic DNA (RAPD), microsatellite or simple sequence repeats (SSR), and amplified fragment length polymorphism (AFLP) . This report summarizes the results of our efforts to generate useful induced mutants of the Philippine banana cultivars, 'Latundan' and 'Lakatan' through irradiation, and to evaluate the usefulness of DNA marker techniques, such as RAPD, microsatellites or SSR, and AFLP, to characterize the genomic alterations in induced mutants of the two Philippine banana cultivars.

GENERATION OF INDUCED MUTANTS OF PHILIPPINE BANANA CULTIVARS BY GAMMA AND FAST NEUTRON IRRADIATION

Induced mutation (either by chemicals or irradiation) has been established as a tool to generate variation in a number of seed- and vegetatively propagated crops. The project successfully adapted established protocols for mutation induction and subsequent propagation of *in vitro* cultured shoot tips in two Philippine banana cultivars.

In vitro culture, radiosensitivity and mutation induction

In vitro cultures of the cultivar 'Lakatan' were provided by the IPB *in vitro* germplasm collection for banana in 1995. Cultures were maintained on modified MS medium supplemented with 20 mM BAP, 40 mg/l cysteine hydrochloride and 40 g/l sucrose at pH 5.6-5.8. Cultures were placed on a horizontal shaker (60 r.p.m.) and continuous proliferation was maintained for 30 days sub-culture intervals in a controlled environment at 28°C and 16 h illumination. Meristem tips containing a meristematic dome of two to three pairs of leaf primordia excised from *in vitro* growing shoots were used for irradiation experiment.

Longitudinally dissected explants were transferred to sterile Petri dishes with a few drops of sterile distilled water and sealed with paraffin. The radiosensitivity assay was conducted using split shoots of the Lakatan irradiated with 10, 20, 30, 40, 50 or 60 Gy gamma rays from a ^{60}Co source in the FAO/IAEA laboratory in Seibersdorf, Austria. There were 12 shoots per treatment. After treatment, split shoot tips were immediately transferred to fresh medium (modified MS, described above) and multiplied to generation

M_1V_4 in proliferating MS medium. Radiosensitivity for gamma radiation of Lakatan was established at 40 Gy. Other workers in this IAEA CRP on banana have obtained similar values (35-45 Gy) using their preferred cultivars such as Embul in Sri Lanka, Pisang Mas in Malaysia, and Paracido al Rey, Gran Enano and Burro CEMSA in Cuba .

After the radiosensitivity test, bulk irradiation of 500 meristems of Lakatan with 40 Gy gamma rays (LK-40) was carried out. Additional irradiation with fast neutron at a dose of 3 Gy was also carried out in 500 and 200 meristems of Lakatan (LK-3) and Latundan (LT-3), respectively. The shoots were sub-cultured to the M_1V_5 as described above. Roots were induced on half-strength MS medium supplemented with 40 mg/l cysteine hydrochloride and 20 g/l sucrose solidified with 7 g/l agar. Plantlets were transferred onto coir dust-soil mixture (1:1) in the greenhouse. After six months, the acclimatized plants were transferred directly to the field for further selection and characterization.

Table 1 Summary of mutation induction in the two Philippine cultivars, "Lakatan" (LK) and "Latundan" (LT) using 40 Gy gamma rays (LK-40) or 3 Gy fast neutrons (LK-3, LT-3)

	LK-40	LK-3	LT-3	Total
Irradiated meristems	500	500	200	1200
M1V5 plantlets produced	4197	2053	307	6557
M1V5 plantlets field planted	533	233	128	894
Mature plants produced	155	30	76	261
BBTV negative plants (ELISA test)	8	12	32	52
Selected clones	5	8	17	30

The results of mutation induction in the two Philippine cultivars used are summarized in Table 1. Irradiation with gamma rays or fast neutrons induced the production of agromorphological variant phenotypes in Lakatan and Latundan cultivars, as described in the following section, and produced clones which gave rise to suckers that remain negative (resistant) to the banana bunchy top virus (BBTV) test. Nevertheless, large number of clones was lost due to BBTV infection, particularly in the mutated populations of the cultivar Lakatan.

Morphological Characteristics of Mutated Populations

Morphological characterization of individual clones in each mutated populations (LK-40, LK-3 and LT-3) was carried out using the IPGRI descriptor for banana . Each clone was characterized using characters taken at the vegetative, reproductive and fruiting stages.

Morphological variants were observed for almost all of the traits measured in the three irradiated populations. These include dwarf habit and large fruit size. In general, only a small proportion of variant phenotypes was observed for each trait. Principal component analysis of 27 vegetative traits was conducted. Most of the clones (depicted as large clusters) in each mutated population were morphologically similar to each other. These clones exhibited traits that are typical for non-irradiated Latundan and Lakatan. More variants were observed in populations irradiated with fast neutrons (LK-3 and LT-3) than with 40 Gy gamma rays. These results support previous observations that fast neutrons result in a higher frequency of induced mutations than gamma rays. Fast neutrons, a particle radiation, are more energetic than gamma rays, a form of electromagnetic radiation .

ESTABLISHMENT OF PROTOCOLS AND IDENTIFICATION OF MOLECULAR MARKERS FOR GENOME ANALYSIS OF INDUCED MUTATIONS OF BANANA

During the past two decades, various DNA marker techniques have been developed. These techniques can be classified primarily as restriction enzyme-based, polymerase chain reaction (PCR)-based, or a combination of both. Of these techniques, RFLP, RAPD, SSR and AFLP have been used in various aspects of genome analysis of *Musa* species. This section will present the results obtained by the project in developing protocols for RAPD, SSR and AFLP analysis of induced mutant populations of two Philippine banana cultivars.

DNA Extraction

The micro extraction procedure of Fulton et al. was adopted for all molecular marker analyses described in the following sections because it is fast, simple and yields relatively clean DNA in sufficient quantity for PCR amplification. This procedure was used to extract DNA from young leaf spindles of normal and irradiated banana clones.

Random Amplified Polymorphic DNA (RAPD)

Random amplified polymorphic DNA (RAPD) is a novel technique developed by Williams et al. and Welsh and McCleland based on the amplification of random DNA segments with single primers of arbitrary nucleotide sequence. The technique allows detection of extensive polymorphism because of the unlimited number of arbitrary primers. The assay is fast, non-radioactive, and requires only nanogram quantities of DNA. The literature is full of reports on the successful applications of RAPD analysis for various genome studies, including genotype identification, in a broad range of species.

The results obtained in this project indicate that the RAPD technique can be readily adapted for banana. A RAPD protocol was adapted from the procedure of Williams et al. with the following modifications in the concentrations of $MgCl_2$ (1.5-2.0 mM), RAPD primer (0.15-0.18 μM), and template DNA (3-5 ng). RAPD primers were purchased from Operon Technology, USA. Amplification was performed using an MJ Research PTC-100 thermal cycler with a temperature cycle profile consisting of the following: 93°C (1 min), 45 cycles of 93°C (30 sec), 36°C (1 min), 72°C (1.5 min), and final extension at 72°C (2 min). Amplification products were separated in 2% agarose gel in 1 × TAE buffer and stained with ethidium bromide. Figure 2 illustrates the RAPD patterns obtained following the procedure described above. Preliminary RAPD analysis was conducted in four banana clones (one normal clone of Lakatan, and two irradiated with 3 Gy fast neutrons and 40 Gy gamma rays; and one clone of Latundan irradiated with 3 Gy fast neutrons). Twelve random primers (OPQ-01 to -03, OPR-01 to -03, OPS-01 to -03 and OPT-01 to -03) were tested. All of the 12 primers tested generated well resolved amplification products in all the four clones tested. Four primers revealed variation among the clones. These primers are OPR-01, OPR-02, OPR-03 and OPT-01. Four RAPD markers (OPR-01_{3832}, OPR-03_{741} and OPT-01_{2600}) were detected only in the Lakatan clone irradiated with 3 Gy fast neutrons. The results showed that all 12 primers were able to differentiate the cultivar Lakatan from Latundan. The amplification patterns produced by Lakatan clones were found to be distinct from the Latundan clone. Lakatan belongs to the *acuminata* cultivars (AAA genome) while Latundan belongs to the *acuminata* × *balbisiana* hybrids (AAB genome).

A similar experiment using 60 RAPD primers yielded similar results: 33% of the primers tested detected variation between irradiated clones, and more differences in the RAPD bands were detected with the clones irradiated with fast neutrons. The results indicated that more point mutations were generated by the fast neutron treatment than by the gamma irradiation in these materials. Despite the ease and discriminating capability exhibited by RAPD markers, problems of reproducibility and transferability were often encountered during the experiments. Hence other marker techniques were considered, such as SSR and AFLP, described in the following sections.

Simple Sequence Repeats (SSRs) or Microsatellites

Microsatellites are another type of PCR-based marker that consist of short DNA sequences (usually 1-6 bp in length) that are tandemly repeated from two to several thousand times. They are also known as simple sequence repeats (SSRs), short tandem repeats (STR), simple sequence length polymorphisms (SSLP), or sequence-tagged microsatellite sites (STMS). SSRs are an important class of DNA marker because they are abundant, uniformly dispersed in the genome, PCR-based, and are highly informative because they are co-dominant markers . SSR polymorphism results from differences in the number of repeat units between individuals at a particular SSR locus. The variation in the number of repeat units is thought to be due to unequal crossing-over or slippage of DNA polymerase during replication of repeat tracts. The microsatellite loci were found to be abundant and highly variable in many eukaryotic DNA examined. Since the DNA sequences that flank the SSR loci are conserved, suitable primers to amplify the SSR loci were designed, resulting in highly specific PCR amplification and therefore high reproducibility. The technology is also readily transferable since information can be communicated as primer sequences.

Our participation in the banana CRP allowed us access to STMS primers and protocols from the University of Frankfurt, Germany. The results obtained show that the non-radioactive SSR marker protocol has been successfully adapted for use with banana. This SSR protocol is highly reproducible and easy to transfer to less sophisticated laboratories such as those in the Philippines.

Genomic DNA was isolated from 1-3 g of fresh leaf tissue from field-grown plants, following the protocol of Fulton et al. PCR amplification was carried out following the conditions described by Kaemmer and co-workers with some modifications. Each 20 μl reaction contained 30 ng of template DNA, 1 × PCR buffer (10 mM Tris-HCl, pH 8.3, 50 mM KCl), 1.5 mM $MgCl_2$, 0.15 mM dNTPs, and 0.2 μM each of forward and reverse primers at 0.4 U of Taq DNA polymerase. Amplification was carried out using an MJ Research thermal cycler using the following profile: initial denaturation, 94°C for 4 min, 35 cycles of 94°C for 30 s, 50-60°C for 30 s and 72°C for 30 s, and a final extension at 72°C for 10 min. Soaking temperature was 4°C. Modification of the annealing temperature for each primer resulted in improved band resolution. Each reaction was overlaid with mineral oil. Bands were resolved using 5% denaturing polyacrylamide gels. A non-radioactive silver staining protocol was used to detect the AFLP fingerprints instead of the recommended g-^{33}P or g-^{32}P radioisotopes for radioactive detection. Figure 3 illustrates the typical SSR banding patterns obtained following the procedure described above.

The SSR polymorphism survey was conducted using DNA extracted from representative clones of one non-irradiated and three irradiated banana populations (LK-40, LK-3, LT-3). The laboratory of Dr. Günter KAHL, University of Frankfurt, Germany, kindly provided a total of 12 *Musa* STMS primer pairs. Twenty-four additional *Musa* SSR primer pairs were purchased from Research Genetics, USA. The results of the polymorphism survey are summarized in Table 2. Of the 36 primer pairs tested, 72.2% (26/36) yielded amplification products from all clones used. All the SSR primers that yielded amplification products could discriminate between the two cultivars tested. However, none of the SSR primer pairs tested, except for Ma1/3, were able to detect variation between the non-irradiated and irradiated clones, or between the clones irradiated with 3 Gy fast neutrons and 40 Gy gamma rays.

Amplified Fragment Length Polymorphism (AFLP)

Amplified fragment length polymorphism (AFLP) is the newest marker technology developed for DNA fingerprinting. AFLP combines the reliability of classical restriction-based fingerprinting methods such as RFLP with the speed and convenience of PCR-based marker techniques. However, there are some major limitations to the adoption of this promising marker technology. First, AFLP is a proprietary technology and is therefore more expensive than other PCR-based marker techniques. Secondly, the AFLP protocol developed by Keygene is based on radioactive detection using radioisotopes, and as such can be used only in laboratories where facilities for radioisotope handling and containment are available.

A method was developed for non-radioactive detection of AFLP fingerprints in banana, suitable for laboratories in developing countries. Ten AFLP primer pairs (+3/+3) were initially tested in five representative samples of irradiated clones. Optimization was based on the AFLP protocol of Vos et al. , AFLP Analysis System I kit (Life Technologies, Cat. No.10544-013) and Promega silver staining kit . The following modifications and additional steps were taken, which cut the cost and ensured high gel/band resolution in a non-radioactive detection system:

- High quality plant genomic DNA was obtained with an additional DNA purification step using phenol instead of the recommended GIBCO-BRL Plant DNAzol. The isolation of good quality intact DNA ensures complete digestion of the genomic DNA.
- Prior to restriction digestion using *Eco* RI/*Mse* I restriction enzymes, digestibility of the genomic DNA samples was tested using cheaper restriction enzymes such as *Eco* R1 or *Hin*d III.
- The reaction volumes for the restriction enzyme digestion, ligation, pre-amplification and selective amplification were reduced to 50% of the recommended total reaction volume for each step without reducing the quality of bands or the efficiency of the reaction. This modification reduces the cost and doubles the number of samples that can be analyzed per kit.
- Dilution of the pre-amplification product was varied from 1:10 to 1:50 depending on the intensity of the pre-amplification product obtained using standard gel electrophoresis on representative samples.
- A non-radioactive silver staining protocol based on Promega was used to detect the AFLP fingerprints instead of the recommended g-^{33}P or g-^{32}P radioisotopes for radioactive detection

AFLP fingerprints of representative induced mutated clones of banana and accessions of *Musa textilis* are shown in Figure 4. The results of the AFLP screening are summarized in Table 3. Eight of the 12 AFLP primer pairs (+3/+3) used could differentiate between the cultivars Lakatan and Latundan, as well as among the different irradiated Latundan clones. The number of bands generated ranged from 18 to 40 (mean, 34 bands) and percentage polymorphism ranged from 31.4 to 77.8%, with a mean of 51.1%.

The results clearly indicated that AFLP is ideal and more useful for fingerprinting purposes than other marker systems because of its high multiplex ratio, i.e. more bands (loci) per gel can be resolved.

While more primer combinations need to be tested, these preliminary results suggest the potential value of this technique for detecting genome variation between cultivars, and for detecting genome alterations in induced mutants of banana, including those showing very similar phenotypes.

Table: Summary of polymorphic AFLP markers in five irradiated Lakatan and Latundan clones

Selective primer pairs	Number of bands	Number of polymorphic bands	% Polymorphism
E-ACG/M-CTC	35	11	31.4
E-ACG/M-CTG	32	20	62.5
E-ACG/M-CTT	18	14	77.8
E-AGC/M-CAA	40	20	50.0
E-AGC/M-CÁC	34	14	41.2
E-AGC/M-CAG	29	14	48.3
E-AGC/M-CAT	31	14	45.2
E-AGC/M-CTA	21	7	33.3
E-AGC/M-CTC	29	19	65.5
E-AGC/M-CTG	32	18	56.2
Mean	34	15.1	51.1

AFLP ANALYSIS OF PARENT CLONES AND SUCKERS OF SELECTED INDUCED MUTANTS

The mutation technique, in combination with *in vitro* micropropagation, is a powerful technique which may well produce desired variation in well established clones. Repeated *in vitro* propagation is incorporated in the procedure to generate the large number of samples required and to dissociate chimeras. In general, mutated cells are difficult to monitor. Induced mutants are usually differentiated from each other and from the original cultivar by phenotypic analysis, which can be severely limited by the large size of the mutagenized population and by dominance, and developmental and environmental effects in the traits used in characterizing the population.

The use of molecular markers may overcome this limitation. Based on the results obtained in this study, and those reported by other workers , AFLP has proved to have the highest potential for detection of genome alterations in somaclonal variants and radiation-induced mutants.

The materials used in this study were generated as described in Section 2. Selected parent clones consisted of those which were morphologically similar to and those with obvious differences from the non-irradiated clones of each cultivar. From 30 selected parent clones of the three mutated populations, LK-40, LK-3, and LT-3, 78 suckers were derived and established in the field. Suckers of non-irradiated Lakatan and Latundan cultivars were planted alongside as controls.

AFLP analysis was carried out following the procedures described in the preceding sections. Six out of 24 primer pair combinations tested namely, E-AGC/M-CAA, E-AGC/M-CAG, E-AGC/M-CAT, E-AAC/M-CAC, E-AAC/M-CAG, and E-AAC/M-CAT, were employed.

AFLP bands were scored manually as present (1) or absent (0). Jaccard's similarity coefficients were computed between genotypes in each clone group, namely primary and secondary suckers of irradiated and non-irradiated Latundan, Lakatan 40 and Lakatan 3, using SAS version 6.12 .

Cluster analysis was applied on the similarity matrix with the unweighted pair group method with arithmetic averages (UPGMA) strategy. Dendrograms were constructed using genetics software MEGA version 2.1.

AFLP Analysis Among Parental Clones

The six AFLP primer combinations generated consistent amplified products, with an average number of 20 polymorphic bands per primer pair. The analysis of 30 parental clones, plus the non-irradiated control, using the six AFLP primer pairs allowed us to calculate the matrix for all pairwise comparisons between the parent clones.

The relationships among the clones in each mutated population, LT-3, LK-40 and LK-3, are illustrated by the dendrograms in Figure 5. The analysis showed that all mutated clones differed from the non-irradiated genotype and between each other in varying degrees. Of the three mutated populations generated, the clones from the 40 Gy (LK-40) gamma-irradiated populations showed the least variation, while both populations irradiated with 3 Gy fast neutrons were more variable. The results indicate that fast neutron irradiation gives rise to more mutations and that four cycles of subculture were not sufficient to eliminate chimeras.

AFLP Analysis between Parent Clones and Derived Suckers

The relationships were also calculated between parent and suckers in each mutated populations, LT-3, LK- 40 and LK-3, as illustrated by the dendrograms and electrophoretograms in Figure 6. The analysis showed that the majority in suckers were different from the parental clones validating further the results that fast neutron irradiation gives rise to more mutations and that the shoot-tip culture technique and the number of propagation cycles employed were not sufficient to eliminate chimeras.

CONCLUSION

The study has produced promising induced mutants of the two most important Philippine banana cultivars, and has established molecular marker techniques for the more efficient characterization and analysis of mutation-induced changes in the *Musa* genome. All the methods developed can be easily adopted in less developed countries where bananas are widely grown. The non-radioactive silver staining technique for SSR and AFLP markers are suitable for local laboratory conditions. The results obtained from both morphological and molecular analyses show that more point mutations were generated by irradiation with fast neutrons than with gamma rays. However, the results indicate that the number of vegetative propagation cycles for the shoot-tip technique used may not be sufficient to eliminate chimeras completely from the mutated populations. Hence, new mutation breeding strategies need to be developed. The results obtained represent significant steps that could provide a sound basis for the successful application of mutation and molecular marker techniques to the improvement of the banana in the Philippines.

13

Biotechnology and in Vitro Mutagenesis

Banana is the second most commonly grown fruit crop in Malaysia. Overall banana production has decreased due to the increasing threat of Fusarium wilt disease, high labour costs and marketing issues. This program was initiated to improve banana cultivars by induced mutations and biotechnology, especially to produce mutant varieties with improved traits such as Fusarium tolerance, short stature plants, early fruiting and high bunch weight. Banana shoot-tip cultures were most suitable for micropropagation for large-scale plant production. Commercial companies have adopted this method, and they produce 1.3 million plants annually, with approximately 0.5% somaclonal variation.

However, the cost of production of *in vitro* plants could be reduced by low-cost micropropagation. Somaclonal variation has been effective in banana for the selection of useful somaclones, e.g. early flowering and tolerance to Fusarium wilt. In Novaria, an early flowering mutant, 7% of the plants survived in the Fusarium 'hot spot' for 3 years. Consequently, somaclonal variation is being used as a strategy to select useful mutants. Since bananas are mostly sterile polyploids, highly heterozygous, and propagated vegetatively, genetic improvement by cross-breeding is an insurmountable task. On the other hand, mutation techniques are highly suitable for banana improvement.

Pisang Berrangan (AAA), a popular dessert banana variety, was gamma-irradiated at several dosages (20-60 Gy). The highest percentage of variants, such as changes in leaf coloration and leaf texture, leaf deformation, stunted growth etc., was obtained with 45 Gy treatment. The double-tray system was developed for the selection of mutants tolerant to Fusarium wilt. Among selected plants showing tolerance to Fusarium wilt, none survived field evaluation under Fusarium 'hot spot' conditions. Molecular characterization with RAPD of resistant and susceptible banana types showed random variation for different markers. However, four primers showed bands specific to either resistant or susceptible seed progenies, but could not provide information on the degree of co-dominance.

INTRODUCTION

Bananas are grown in 122 countries, with a cultivated area of 3.8 million hectares and a total production of 56.4 million metric tonnes. In Malaysia, banana is the second most widely cultivated fruit, covering about 26,000 ha with a total production of 530,000 metric tonnes. About 50% of the banana growing land is cultivated with Pisang Berangan and the Cavendish type, and the remaining popular cultivars are Pisang Mas, Pisang Rastali, Pisang Raja, Pisang Awak, Pisang Abu, Pisang Nangka and Pisang Tanduk. Bananas are cultivated for local consumption by smallholders, and only about 12% of the total production is exported, mainly to Singapore, Brunei, Hong Kong and the Middle East. However, banana production in Malaysia has decreased because of an increasing threat of diseases (particularly Fusarium wilt), high labour costs and marketing issues.

The present research program was initiated with the objective of improving the important dessert bananas in Malaysia. This includes production of resistance or tolerance to Fusarium wilt or Panama disease, short plant stature, early fruiting, and high bunch weight. Banana cultivars are vegetatively propagated clones and are generally triploids and sterile. Tissue culture techniques have been exploited for (a) propagation of selected lines or natural variants; (b) generation of somaclones; (c) production of meristem pieces for *in vitro* mutagenesis and polyploidy induction; and (d) zygotic embryo culture to generate seed progenies for genetic and molecular studies.

MICROPROPAGATION OF BANANAS

Ma and Shi reported *in vitro* culturc of banana. Shoot-tip culture in combination with heat therapy was successfully used to produce virus-free banana plants as well as for mutation induction. Several reviews have been written on the *in vitro* culture techniques for different applications.

The shoot-tip meristem culture technique is used routinely for large-scale propagation of bananas in Malaysia . Currently, commercial production of micropropagated banana plants in Malaysia is estimated at around 1.3 million. The commercial micropropagation protocol has been successfully modified to bring somaclonal variation down to 0.5% . There are several advantages of tissue culture-derived banana plants, such as vigorous growth, high survival rate, uniformity, and high flowering rate (90% of plants flower compared with 59% of plants produced from suckers). This shortens the harvesting period by about 2 months. Reduction in the cost of micropropagation has been achieved by using locally made culture containers, and by replacing tissue-culture grade sucrose with commercial cane sugar, and Gelrite with locally available agar. We also found that the multiplication rate could be enhanced by 33% by using liquid or semi-liquid media instead of solid medium. We could also achieve rooting by placing plantlets in a hydroponic system in the greenhouse; the resulting rooted plantlets needed a shorter period of hardening. Other cost reduction strategies, particularly the use of sunlight, should be pursued to reduce the current production cost further.

EXPLOITATION OF SOMACLONAL VARIATION

The occurrence of off-type plants, commonly known as somaclonal variants , is relatively common in *in vitro* plants. Usually, most of the somaclones are undesirable from agronomic point of view. Nevertheless, the phenomenon is useful in generating genetic variation for banana improvement:

(a) With continuous selection of somaclones for earliness from tissue culture-derived commercial plantings of Pisang Berangan, the proportion of plants showing earliness in flowering increased gradually from 16% to over 60% at 8 months of planting (Table 1). This means that at about 11 months after planting, 66% plants were ready for harvest compared with 4% for sucker material and 16% for unselected tissue-cultured plants.

Table: Selection of somaclones for early flowering in Pisang Berangan

			% plants with flower emergence (months after field planting)				
Year	Material	No. planted	5	6	7	8	Total
1990	Sucker	48	0	0	0	4	4
1990	TC	619	0	0	0	16	16
1991	TC	77,000	0	1	5	10	16
1992	TC	60,000	1	7	23	32	63
1996	TC	19,000	1	8	24	33	66

(b) Using tissue-culture-derived plants, 27% of Pisang Rastali (one clone) survived for more than 3 years in the Fusarium wilt-infested field, while all other clones planted at the same time had succumbed to the disease. Dissection showed the symptoms of Fusarium infection at the base of the corm tissue. The vascular discoloration was arrested 20-30 cm above the base. The continuous selection of tolerant plants has resulted in the release of a Fusarium wilt-tolerant clone of Pisang Rastali called Mutiara :

1990-1993: 27% of the micropropagated plants from one Pisang Rastali clone survived for more than 3 years in the Fusarium-infested field.

1994 planting: Rate of survival at one year after planting was 51%.

1995 planting: Rate of survival at one year after planting was 91%.

1996 planting: Rate of survival at one year after planting was 100% and 80% at 3 years after planting.

Table: Survival of Pisang Rastali in a Fusarium 'hot-spot'

Cultivar	Number planted	% surviving plants at months after planting								
		2	4	6	8	10	12	18	24	36
Pisang Rastali	60	100	100	100	100	100	100	85	70	60
Gold Fingers	60	100	100	100	100	100	100	85	85	80
Novaria	60	72	53	37	35	12	12	7	7	7
Pisang Berangan	60	63	33	4	0	0	0	0	0	0

Table: Screening somaclonal variants in the Fusarium 'hot spot' for disease tolerance

Cultivar	Number planted	% surviving plants at months after planting								
		2	4	6	8	10	12	18	24	30
Novaria-VBB	240	79	57	35	15	6	3	2	2	0
Novaria	688	100	97	69	47	13	9	9	9	0
Novaria	1454	96	86	74	45	15	3	1	0	0
Novaria V53-3a	4	100	75	50	25	25	25	25	25	25
Novaria-V53-4a	29	69	24	3	0	0	0	0	0	0
Novaria1	14	100	100	43	29	29	29			
Novaria2	24	100	88	0	0	0	0			
Novaria3	21	100	90	33	24	5	5			
Novaria4	47	100	96	32	9	4	4			
Novaria5	67	100	94	22	6	0	0			
Berangan	850	100	100	63	6	0	0			
Berangan	586	100	100	90	36	0	0			
Berangan	130	100	100	78	54	13	4			

Novaria, an early flowering mutant of 'Grande Naine', was one of the susceptible controls used. It was encouraging to note that 7% of the plants survived in the Fusarium 'hot spot' for 3 years. Consequently, exploitation of somaclonal variation has been adopted as a regular strategy to select for Fusarium-wilt resistance in Novaria and Pisang Berangan. It was also shown that lines derived from different suckers of the same mother plant tended to show variation in degree of tolerance.

Table: Line variation in response to Fusarium wilt tolerance in the 'hot spot'

Selection	Number planted	% surviving plants at months after planting				
		1	2	3	4	5
Berangan	20	95	55	5	0	0
Berangan IFTA-1	20	100	100	80	60	25
Berangan IFTB-2	20	100	75	50	10	10
Novaria-1	70	100	93	26	6	3
N199A-1	20	100	90	70	50	45
N199A-2	45	100	96	56	33	27
N199A-3	13	100	100	93	93	85
N299A-1	20	100	85	55	35	35
N299A-2	31	97	94	58	46	39
N299A-3	65	100	100	67	42	40
N199F-1	20	100	75	30	25	15
N199F-2	16	100	82	25	13	13
N199G-1	20	100	90	25	10	5
N199G-2	16	100	82	19	7	0
N199N-1	20	100	85	35	30	20
N199N-2	16	100	100	32	7	7
N199Q-1	20	100	95	55	50	35
N199Q-2	18	78	6	6	6	0
N199R-1	20	100	85	50	35	30
N199S-2	16	100	82	19	0	0
N199T-1	20	100	80	50	45	45
N199T-2	18	100	78	39	17	6
Gold Finger	20	100	100	100	100	100

IN VITRO MUTATION BREEDING

Edible bananas are mostly sterile polyploids and must be propagated vegetatively, hence genetic improvement through cross-breeding is not possible. Mutation breeding has been suggested as an excellent alternative approach for banana improvement. In addition, the heterozygosity of asexual banana clones makes them suitable for mutation induction. The heterozygotic status is expected to be Aa in loci of diploid cultivars while the triploids of A genomic types can exist in either Aaa or Aaa forms. For interspecific hybrids, the heterozygotic constitution could be AaB, Aab, AAb, ABb, aBb or aBB. Mutation induction may uncover a recessive phenotype by mutating, inhibiting or deleting the corresponding dominant allele .

The use of cultured shoot tips for mutagenesis has facilitated mutation induction and the regeneration of potential mutants. An early flowering mutant of Grand Naine, GN-60Gy A, which also showed differences in the zymograms of soluble proteins and esterase isozymes, was induced by Novak after exposing shoot tips to gamma radiation. Further selection of GN-60Gy A in Malaysia has resulted in the release of an early fruiting Cavendish banana called Novaria . Matsumoto and Yamaguchi selected an aluminium-tolerant mutant from irradiated protocorm of a Cavendish banana.

Pisang Berangan (AAA) is a popular dessert banana, having good fruit quality, flavour, colour, pulp texture, size and shelf life. However, it is relatively tall and very susceptible to Fusarium wilt (*Fusarium oxysporium f.*sp *cubense*) and freckle disease caused by *Cladosporium musae.* Consequently, a mutation breeding program was initiated using gamma irradiation to induce genetic variation so that plants could be selected with one or more of the following characteristics: (a) tolerance to Panama disease; (b) short plant stature; and (c) early fruiting and high bunch weight. The present review aims to present the current status of the research, problems encountered, and research strategies.

Techniques for *in vitro* Mutation Induction

Healthy and normal plants from selected Pisang Berangan were micropropagated to produce shoots/buds with apical domes from which meristem pieces (about 1 cm × 2 mm) were aseptically excised. About 20 to 30 meristem pieces were transferred to each sterile Petri dish, which contained a piece of sterile moist paper, and then sealed with parafilm. The shoot-tip meristems in the Petri dishes were irradiated in a gamma cell with a ^{60}Co source to give the required doses.

The treated explants were washed thoroughly with sterile distilled water prior to culture on modified MS medium supplemented with 150 ml/l coconut water, 30 g/l sucrose, 5 mg/l BA and 5 g/l agar (A02 Type 900). Subsequently, each growing shoot was separately cultured to generation M_1V_4 (three subcultures at monthly interval) and then allowed to root (3 to 4 weeks). As mutation is a single cell event, mutation induction in a multi-cellular shoot-tip would result generally a chimera in the M_1 plants. Hence, mutated cells may be recovered through repeated *in vitro* propagation of mutated cell progenies in several vegetative cycles. The rooted plantlets were transplanted to soil mix in perforated polythene bags [5" × 7" (approx. 12.5 × 18 cm)] for acclimatization (8-10 weeks) before field planting at 2.4 × 24 m. Evaluation of mutagenic effects was conducted on both nursery and field-grown plants.

Dose Sensitivity and Mutagenic Effects

A pilot experiment was conducted using doses of 0, 25, 35,45 and 60 Gy of gamma irradiation to study radiation sensitivity of Pisang Berangan . Both survival and shoot multiplication rate decreased with an increase of radiation dose (Table 5). The LD_{50} dose (the radiation dose that reduces the growth to 50% of that of the control) for Pisang Berangan was 38 Gy. This is in agreement with Novak, who found LD_{50} doses of 25 Gy (2n plants), 35 Gy (3n plants, AAA), 40 Gy (AAB and ABB plants), and 50 Gy (4n plants, AAAA). Whether LD_{50} dose is a useful criterion for mutation induction is debatable. According to the available information, the doses being used currently range from 20 to 60 Gy of gamma irradiation. The general recommendation is to use a low dose on a very large number of primary explants (e.g. 10,000) followed by one subculture *in vitro*, or a high dose on a large number of primary explants (e.g. 2000) followed by at least three subcultures to M_1V_4. The present radiosensitivity results indicated that the highest percentage of variants was obtained at 45 Gy, but that at 60 Gy many lethal or abnormal mutants would probably not have survived.

Table: Survival and multiplication rate per explant at 12 weeks after gamma irradiation

Dose (Gy)	0	25	35	45	60
Number of explants	477	438	462	532	507
Survival %	73	66	48	48	28
Multiplication rate [a]	4.35	3.30	2.35	1.90	0.65

[a] No. of buds/shoots produced/explant

Table: Variability induced by gamma irradiation of Pisang Berangan

Dose (Gy)	0	25	35	45	60
% of variants (nursery plants)	2.9	11.2	14.7	16.2	10.5
% of variants (field-grown plants)	5.4	20.1	21.5	16.2	10.5
Weeks to harvest:					
Mean (sample size)	49.3 (228)	50.3 (181)	49.4 (174)	48.9 (153)	48.6 (108)
CV (%)	5.5	8.4	9.1	9.2	9.4
Early fruiting (%)[a]	2.2	2.8	4.6	9.2	22.2
Height at flowering:					
Mean (sample size)	305 (232)	312 (179)	295 (208)	292 (174)	291 (114)
CV (%)	8.0	9.9	12.5	11.1	12.8
Early fruiting (%)[b]	2.3	11.2	14.9	14.4	10.5

[a] % of plants harvested at 8 weeks earlier than the mean

[b] % of plants with 250 cm high or less

Mutagenic treatments caused many morphological changes in both nursery and field-grown plants. These phenotypic variants were increased 4- to 6-fold by gamma irradiation (Table 6). Some of these variants include:

- o Changes in leaf coloration which included discoloration, darker green, and various forms of leaf variegation;
- o Leaf deformation, such as uneven lamina; leaf becoming smaller, narrower, long but weak, broad and oval; tendency to roll up at edges;
- o Changes in leaf texture such as waxy, leathery or rough, crumpling and rough veins;
- o Pseudostem becoming flat or stiff or bulbous;
- o Various degrees of stunted growth.

Additional phenotypic variants observed in field-grown plants include:

(a) Stunted growth or choking appearance;

(b) More compact and erect leaves;

(c) Pseudostem becoming flat, stiff or even split and its colour varying from dark brownish (more anthocyanin) to lighter yellow with waxy petiole;

(d) Various forms of deformation in bunch and fruit characteristics.

For the two important agronomic traits, namely number of weeks to harvest and plant height at flowering, the mean values of irradiated plants did not differ significantly from the control plants. However, mutation induction tended not only to increase the variability, as indicated by the increase in the coefficient of variation (CV), but also showed a much higher frequency for plants having early fruiting or shorter height. The amount of variability observed for these two quantitative traits may be underestimated, because many mutagenised plants might not survive to maturity due to the adverse effects of mutagenic treatments.

The results have demonstrated that mutation breeding coupled with *in vitro* technique could produce morphological changes as well as an increase of variability in quantitative traits. Though many morphological changes are undesirable, this source of induced variation is potentially useful for the sterile *Musa* species.

Screening for Desirable Traits

Screening for agronomic traits

As mutation is a random and rare event and mutants are generally recessive and deleterious, mutation breeding requires the screening of a large sample of mutagen-treated populations to identify the desired individuals. Screening has to be considered at the different stages of growth development. Field screening of large populations, particularly for agronomic traits, was impractical in view of the space required, the management needed, and the 2-year cycle needed for selection. Phenotypic selection of M_1 plants for important agronomic traits, such as early fruiting, shorter plant stature and high bunch weight, was unreliable because the minor variations among plants are confounded with the deleterious effects of mutagens and with environmental factors such as soil fertility, water stress and management practices.

Screening for disease tolerance in mutated materials

Panama disease caused by the soil-borne fungus *Fusarium oxysporium* f.sp. *cubense* is an important factor limiting commercial cultivation of bananas in Malaysia. Control measures involving field sanitation, soil fumigation and liming are not effective. Although the fungal infection starts with invasion of the roots before ingress into the vascular system, the expression of visible symptoms results from occlusion by the fungus mycelium, induction of tyloses, enzymatic breakdown of parenchymatous tissues and the action of fusaric acid, a potent mycotoxin, produced by the fungus. Mechanisms of resistance are believed to be multi-faceted, and plant age and vigour are contributing factors.

To facilitate disease screening, a quarantine field heavily infested with the pathogen was established as a Fusarium 'hot spot' in an oil palm plantation. This facility allowed complete disease screening at nursery and field planting stages. Eight- to ten-week-old nursery plants derived from gamma-irradiated materials were planted in the Fusarium 'hot spot' to screen for disease resistance. In the early pilot experiment to study radiation sensitivity, 2000 plants were planted in a field which was later heavily infected with Fusarium wilt. All plants eventually died except one, which survived for 2 years in the field. This plant, derived from 45 Gy gamma-ray treatment, was designated as I289. Two suckers from this selection were micropropagated to produce plants for further confirmatory testing against the Fusarium pathogen. However, the results indicated that the tolerance of this selection had broken down, as the survival rates were only slightly higher at different stages of assessment those of the control. This may indicate that I289 was merely an 'escape', or that the pathogen population in the 'hot-spot' is too high to allow the survival of any tolerant plants.

Table: The survival of plants derived from I289 in the Fusarium 'hot spot'

Selection	Number planted	Number of plants surviving at months after field planting					
		1	2	3	4	5	6
I289	727	527	227	156	61	22	0
Control	167	162	145	97	41	3	0

As tolerance to Fusarium wilt appears to show quantitative variation, an attempt was made to irradiate *in vitro* meristems derived from I289 at 40 Gy with the hope of increasing the degree of tolerance by accumulating mutagenic changes of minor genes for Fusarium wilt tolerance. A total of 49 plantlets were produced and planted in the Fusarium 'hot spot' in 1996. A few plants showed mild leaf yellowing after 2 months, but all succumbed to the disease eventually. In another experiment, 4000 plants derived from 35 Gy gamma irradiation were planted in a disease-free area; 1818 suckers were taken from normal fruit-bearing plants and tested in the Fusarium 'hot spot'. Five plants were found to survive after one year, but then succumbed to the disease.

From all the screening trials, it was generally observed that plants of 4-5 months old became very susceptible. Susceptible plants generally did not survive for more than 10 months in the field.

Table: The survival of gamma-irradiated Pisang Berangan in the Fusarium 'hot spot'

Dose (Gy)	Number planted	Number surviving at months after field planting				
		2	4	6	8	10
30	423	406	174	0	0	0
40	5143	3288	367	41	19	3
60	2160	1655	606	130	42	0

After screening about 12,000 mutagenic plants, no resistant plants had been identified. Though some selections appeared promising initially, they became susceptible under prolonged testing. It appears that the inoculum concentration in the Fusarium 'hot spot' may be too high to allow the survival of moderately resistant plants. As tolerance to Fusarium wilt appears to show quantitative variation, the 'hot spot' may not be effective in the identification of tolerant lines controlled by minor genes.

Development of techniques for screening resistance to Fusarium wilt disease - the 'double-tray System'

Field evaluation is the most reliable method for disease resistance screening, but is demanding in terms of cost, manpower and space requirements. There is also the need to maintain strict quarantine control to avoid pathogen spread. In addition, plants tend to show symptoms only after 4-5 months. The uneven distribution of pathogen in the field can lead to 'disease escape', while many variables that can affect infection and symptom expression cannot be altered or controlled. After two years of experimental trial, an alternative method of screening seedlings at the nursery stage has been developed. The technique is a modification of the double-cup sand-culture containment method developed earlier for laboratory testing of pathogen virulence. The double cup was replaced by a 'double-tray system'.

The double compartment consists of a tray measuring 43 × 29 × 9 cm that fits snugly into another, larger, outer tray measuring 46 × 31 × 20 cm (Plate 2). The upper tray was three-quarters filled with sterilized fine river sand. The plantlets were watered with tap water every other day plus Hougland's solution once a week.

After about 45 days of acclimatization, the test plantlets were carefully uprooted and only those with healthy white roots were selected for inoculation by immersion in the appropriate conidia suspension for 2 h before being tagged and replanted in the trays for maintenance and observation in the greenhouse.

Table: Response of some banana cultivars to* Fusarium oxysporum *f. sp.* cubense *(FOC)

Cultivars	Genome	Disease Reaction	
		FOC Race 1	FOC Race 4
Pisang Mas	AA	T	S
Pisang Lemak Manis	AA	T	T
Pisang Jari Buaya	AA	R	R
Pisang Berangan	AAA	S	VS
Pisang Embun	AAA	VS	VS
Pisang Udang	AAA	S	S
Grand Naine	AAA	R	S
GCTCV215-1	AAA	R	T (?)
Pisang Serendah	AAA	R	T
Pisang Rastali	AAB	VS	VS
Mutiara (selected P. Rastali)	AAB	T	T
Pisang Seribu	AAB	S	S
Pisang Raja	AAB	S	S
Pisang Relong	AAB	S	S
Pisang Nangka	AAB	S	S
Pisang Awak	ABB	T	S
Pisang Tanduk	ABB	S	S
Pisang Abu Keling	ABB	T	T
Pisang Abu Nipah	ABBB	S	S
Gold Finger	AAAB	R	T

R, resistant; T, tolerant; S, susceptible; VS, very susceptible.

Two-month-old plantlets (10-15 cm height) were found to be suitable for differential expression of symptoms, similarly to field evaluations at the Fusarium 'hot spot'. Seedlings with heights of less than 10 cm did not produce consistent symptoms.

Race 4 of *Fusarium oxysporum* f. sp. *cubense* was freshly isolated from susceptible 'Novaria' plants on Potato Dextrose Agar followed by single spore isolation and inoculum preparation in Armstrong's Liquid Medium. Cultures were incubated at room temperature and shaken twice a day for 7 days, after which they were filtered through two layers of cheesecloth. Inocula at the desired concentrations, comprising mostly microconidia, were then prepared with the aid of a haemocytometer, and immediately used for plantlet-root inoculation. Root immersion for 2 h in the pathogen suspension was found to be effective. All susceptible plants produced foliage and rhizome symptoms at inoculum concentrations higher than 5 × 10^2 spores/ml. All susceptible plants tested, namely Pisang Berangan and Novaria, showed both foliage and rhizome symptoms within a period of 10-28 days. However, more than 94% of tolerant plants (namely Goldfinger and Mutiara), as well as all the uninoculated control plants, did not show any symptoms.

This 'double-tray' technique is being used for preliminary screening of somaclonal variants and induced mutants. Meanwhile, the response of some banana cultivars to *Fusarium oxysporum* f. sp. *cubense* was also noted.

Induction and performance tetraploid Pisang Mas

In Malaysia, Pisang Mas, which belongs to AA group, is well known as one of the most popular local dessert cultivars. Its synonyms are 'Amas' (Philippines) and 'Kluai Khai' (Thailand). The dessert banana has potential for local and export markets. However, it has a low yield (an average of 8-12 kg/bunch), its skin is thin, and its fruit size is small. Therefore, the potential of polyploidy for the improvement of Pisang Mas was studied. The induced tetraploid might also be used for crossing to a selected diploid to produce a secondary triploid.

Shoot meristem pieces obtained from *in vitro* cultured plantlets were immersed in shallow 0 (control), 0.25, 0.5. 0.75, and 1.0 % (w/v) aqueous colchicine solutions for 1, 2, 3 or 4 hours by placing on an orbital shaker (100 r.p.m.) at room temperature (25-28°C). **Sterile** distilled water was used as a control instead of colchicine. After treatment the meristem pieces were washed with sterile distilled water and cultured on semi-solid modified Murashige and Skoog (MS) medium supplemented with 30 g/l sucrose, 0.05 g/l citric acid and 2 mg/l BA, at pH 5.7. Each meristem piece was propagated every 3-4 weeks for three vegetative cycles. Ploidy levels were then determined by flow cytometry, and the tetraploid plants were confirmed by chromosome counts. A high level of tetraploids was induced with 1.0% colchicine treatment for 3 h (41.18%). However, this treatment caused heavy damage and the casualty rate was 27.03%. The treatment that gave high tetraploidy (15%) with low mortality (7.48%) was 0.5% colchicine for 2 h. For all treatment combinations, the rate of tetraploid induction varied from 2.6% to 37.5% and mixoploidy ranged from 14% to 47.6%.

The peak at the left represents debris. Nuclei released from a diploid plantlet appeared at channel 50 for G_1 phase. Plantlets with double the number of chromosomes have their G_1 peak at channel 100, and are therefore tetraploid. Too high a concentration or too long a duration of colchicine treatment might result in higher ploidy levels.

Chimeric plants were produced by even shorter time treatments with low concentrations of colchicine. By prolonging the treatment, more cells in a single meristem can be exposed for chromosome doubling. However, overtreatment may result in a ploidy level higher than tetraploid and increased mortality. Colchicine has direct toxic effects and acts as a strong killing agent even at millimolar concentrations. The chimerism would probably be inherited or gradually revert to diploidy through continuous micropropagation.

The length of the stomata in diploids is about twice that in tetraploids. However, mixoploid plants are intermediate between diploid and tetraploid plants. Breadth measurements show that 22.0% of mixoploids are diploid while 77.0% are tetraploid. However, length measurements show that 4.0% of the mixoploids are diploid and 74.5% are tetraploid. The adaxial leaf surface has significantly fewer stomata than the abaxial. The stomata densities of mixoploids and tetraploids were significantly different from the control ($P < 0.05$). However, the stomata densities on both the adaxial and abaxial surfaces of the diploids and mixoploids overlapped.

The octoploid derived from the doubling of the tetraploid tends to show stunted growth. The octoploid had a big pseudostem and grew very slowly compared with the diploid. The multiplication rates for diploids and tetraploids were 1.02 ± 0.33 and 0.89 ± 0.38 respectively. Slower growth rates and delayed appearance of buds or shoots were also observed in other polyploid studies. After 4 weeks, tetraploid plants have a big pseudostem, shorter stature with a thicker leaf, and are a slightly darker green than diploid plants. Comparisons between diploid and tetraploid plants are shown in Table 10. All parameters measured were significantly different between diploid and tetraploid plants ($P < 0.05$) except the primer root diameter.

Table: Morphology of diploid and tetraploid plants after 4 weeks in culture

	Diploid		Tetraploid		
	Mean	SD	Mean	SD	*t*-value ($P < 0.05$)
Height (mm)	39.97	9.90	24.23	6.75	11.45[a]
Circumference (mm)	12.40	1.92	15.15	1.70	-9.02[a]
Pseudostem diameter (mm)	3.93	0.60	4.78	0.55	-9.10[a]
Primer root diameter (mm)	0.78	0.11	0.80	0.13	-0.68 (ns)
Number of roots	4.59	1.32	4.05	0.87	2.96[a]
Leaf Index (L/W)	2.57	0.91	2.05	0.40	5.40[a]
Leaf length	39.14	11.22	21.39	6.12	11.98[a]
Leaf width	15.22	4.67	10.43	2.53	7.80[a]

ns, non significant at $P < 0.05$

[a] significant at $P < 0.05$

In the field, the tetraploid plants showed drooping leaves and non-compact pseudostems with sucker production. However, the plants tended to be more vigorous and taller, with bigger girth, but smaller bunch size than diploids.

Table: Field performance of tetraploid Pisang Mas

Characters	Diploid P. Mas (AA) ($n = 14$)	Tetraploid P. Mas (AAAA) ($n = 8^a$)
Plant to flowering (months)		
Mean	8.2	7.8
Range	5.5-11.3	7.0-8.3
Height at flowering (cm)		
Mean	234	320
Range	180-275	299-339
Girth at flowering (cm)		
Mean	51	66
Range	38-64	60-69
Bunch weight (kg)		
Mean	5.9	3.8
Range	4-9.2	3-4.6

[a] 50% of the plants remained non-fruiting after 15 months in the field.

Studies on wild banana, Musa acuminata ssp. malaccensis

Malaysia is one of the centres of diversity for both cultivated and wild bananas. Co-evolution of host and pathogen is believed to give rise to the diversity and complexity of Fusarium wilt disease. *Musa acuminata* ssp *malaccensis* (AA) shows good bunch and fruit characteristics. It is highly fertile and the fruit is gummy with plenty of seeds. Research has been undertaken to use seed progenies for variability characterization based on flow cytometric analysis of ploidy and nuclear DNA content and anthocyanins, polyploidy induction and genetic studies on Fusarium wilt resistance.

In vitro zygotic embryo culture

Due to low seed germination under natural soil conditions (2% in a greenhouse study), *in vitro* zygotic embryo culture was developed. Seeds were soaked in 1.4% (v/v) sodium hypochlorite solution for

10 min followed by a quick wash with 70% (v/v) ethanol. Embryos were removed and cultured in glass jars (60 × 80 cm), each jar containing ten embryos. The culture medium consisted of MS salts supplemented with nicotinic acid (0.125 mg/l), ascorbic acid (0.2 mg/l), thiamine HCl (0.5%/l), pyridoxine HCl (0.125 mg/l), myoinositol (2.5 mg/l), glutamine (150 mg/l) and sucrose (5%).

Table: Factors affecting the growth of* in vitro *grown embryos

Treatment	Days to shooting (n = 45)	Days to rooting (n = 45)	Mean stem height (n = 40)	No. of roots/plant (n = 10)	Mean root length (cm) (n = 45)
Lighting					
Light	7.3	9.7	3.0	7.0	3.2
Dark	8.9	7.7	3.9	11.0	5.0
t-value	3.9[a]	2.9[a]	8.9[a]	5.0[a]	6.3[a]
Gelling Agent					
Gelrite	7.6	9.0	4.9	13.0	6.6
Agar/T-900	5.5	6.0	4.9	14.8	10.5
t-value	3.4[a]	2.5[a]	0.8	0.001	2.4[a]
Embryo orientation					
Embryo fully exposed	5.0	7.3	5.3	14.7	11.4
Haustorium embedded	5.3	8.7	3.7	13.1	10.9
Meristem embedded	6.0	10.0	3.1	12.0	10.2
Embryo fully embedded	6.3	14.0	4.0	9.9	9.3
F-value	3.0	11.5[a]	20.6[a]	3.1	2.4

[a] treatment means are significantly different

In vitro embryo culture increased the germination rate to more than 90%. However, there were seeds in which either endosperm or embryos or both were absent. Vuylsteke and Swennen explained that the low germination of mature triploid *Musa* seeds was caused by embryo malformation. The effects of lighting conditions, gelling agent and embryo orientation on some growth conditions were observed for a period of one month. Darkness delayed shoot emergence (8.9 days) as compared to light (7.3 days). Dark-grown embryos produced longer shoots (3.9 cm) and roots (5.0 cm) than light-grown embryos. Similarly embryos grown in the dark produced more roots (11.0) than light-grown embryos (7.0). Raghavan and Torrey reported that 12 h light and dark cycles inhibited root growth. However supplementing the medium with low concentrations of IAA promptly overcomes this inhibition. This may imply that light causes the inactivation of an auxin-like substance necessary for root formation in embryos

Gelling agent had no significant effect on shoot length or number of roots per seedling. Embryos grown on Gelrite produced shoots earlier (5.5 days) compared to agar (7.6 days). A similar trend was observed for root emergence where roots appeared in 6 days on Gelrite compared to 9 days on agar. Roots were significantly longer on Gelrite (10.5 cm) than agar (6.6 cm). These might be related to the presence of impurities in the agar.

Embryo orientation did not have a significant effect on the number of days to shoot emergence, number of roots, or root length. However, long shoots (5.5 cm) were produced by completely exposed embryos. Roots developed much earlier (7 days) with fully exposed embryos, followed by embryos partially embedded (8-10 days) whereas embryos embedded fully in the medium required more time (14 days) to develop roots.

The best morphological characteristics were obtained when the haustorium end touched or was slightly embedded in the media. Afele and De Langhe reported that embryos of *Musa balbisiana* with their longitudinal axis laid flat and half-embedded on solidified agar medium produced the highest germination and the most desirable plantlet characteristics. They obtained a germination rate of 94% *in vitro* within 7 days, but only 50% germination after 54 days for greenhouse-grown seeds.

Genetic studies on Fusarium tolerance

Ten to 15 suckers of *Musa acuminata* ssp. *malaccensis* were collected from five locations and planted in the Fusarium 'hot spot' with Pisang Berangan randomly planted among the tested suckers as a susceptible control. Results showed that all Pisang Berangan plants succumbed to Fusarium wilt disease within 4-5 months, whereas all suckers survived for more than 2 years without any disease symptoms. Dissection of the rhizomes also revealed no disease symptoms.

Table: Response of the wild* Musa acuminata *ssp.* malaccensis *suckers to FOC race 4 after 2 years in the Fusarium 'hot spot'

No.	Samples	Locations	Number of suckers	Response to FOC race 4
1	RI	Kajang, Selangor	15	Resistant
2	IPTJ	University Malaya, Kuala Lumpur	10	Resistant
3	BD1	Sungai Buloh, Selangor	10	Resistant
4	BD2	Damansara Highway, Selangor	15	Resistant
5	BC1	Genting Foot Hill, Pahang	15	Resistant
6	Pisang Berangan	Susceptible control	15	Very susceptible

Seed progenies from the above five populations were produced through *in vitro* zygotic embryo culture and each individual was micropropagated to produce clones for screening Fusarium wilt tolerance using the 'double-tray' system. Seedlings from different populations showed variable response to FOC race 4. Within each progeny, segregation was observed for Fusarium wilt tolerance. Visible symptoms appeared 2-3 weeks after inoculation. Leaf chlorosis started with the older leaves and then progressed upward to the younger ones. Occasionally, a split pseudostem was seen. Plants that showed severe foliar yellowing also had extensive rhizome discoloration. The highly susceptible controls died within 4 weeks.

Table: Seed progenies showing segregation for Fusarium wilt tolerance

Sample	Non inoculated	Resistant reaction (R)	Susceptible reaction (S)	Ratio of R:S
RI	65	19	46	1:3 (ns)
IPTJ	67	23	44	1:3 (ns)
BD2	75	10	65	3:13 (ns)
BC1	54	20	34	1:3 (ns)
BD1	68	56	12	3:1 (ns)
Total	329	128	201	

ns, not significant at 5%

The different genetic ratios observed among five populations suggested that the material used was very diverse genetically. The five populations were obtained from different areas and originated from different stocks of mother plants. Hence they may be different in their genomic compositions and thus their response to FOC. It appears that the seed progenies were very heterogeneous, as wild bananas are both out- and inbreeders. Wild bananas are thought to have coevolved with the FOC pathogen, and their differential response might also be a reflection of the complexity of the pathogen, as indicated by the diversity of the VCG complexes isolated in Malaysia.

RAPD marker characterization of resistant and susceptible seedlings

Leaf samples were collected and DNA extracted from five resistant (R) and five susceptible (S) clonal seedlings of *Musa acuminata* ssp. *malaccensis*. Initially 28 arbitrary primers, 10 bases in length, were evaluated, of which only 15 produced DNA products. Good amplification was obtained with a concentration of 1.5 mM of $MgCl_2$.

A total of 96 major scorable amplification products (bands) from 15 primers were obtained from the ten resistant (R) and susceptible (S) seedlings. Only 10 out of 96 markers were monomorphic and shared among the seed progenies, whereas the remaining 86 markers were polymorphic. The number of scorable RAPD fingerprints generated per primer varied from 4 to 9, with an average of 6 markers per primer. In general, the amplified DNA fragments ranged from 200 bp to 1.5 kbp.

The variation observed for different markers was randomly present in both resistant and susceptible seed progenies. The markers specific to resistant and susceptible seed progenies were only partially resolved by most of the primers used. However, there were four primers that showed bands specific to either resistant and susceptible seed progenies. Seed progenies amplified with the Primer OPA-03 showed a major band of 500 bp shared by all susceptible plantlets that was absent in resistant progenies.

A band of 1000 bp derived from the Primer 21 was observed to be present only in the resistant seed progenies but absent in the susceptible ones. In the case of Primer 27, a band of 1.5 kbp was shared by four of the five susceptible plantlets but was absent in all resistant progenies. For Primer 24, a major band of 400 bp was observed in susceptible progenies but was absent in all resistant plantlets. The remaining primers showed high polymorphism, but without specificity to either resistant or susceptible plantlets. These results demonstrated that RAPD is a robust technique for characterizing seed progenies, but it did not provide any information about the degree of co-dominance. Hence, they may be of low utility for the study of segregating populations where information about heterozygous individuals is very vital. Furthermore, the repeatability of these RAPD results needs to be established.

14

Plant Improvement through Gamma Irradiation

Banana is the most widely consumed fruit in Sri Lanka, and is an attractive perennial fruit crop for small farmers. This is due to its high economic gains throughout the year compared with rice. Lowland rice fields have been converted for banana cultivation. Among the local cultivars, Embul (Mysore, AAB) is in the highest demand for cultivation. Since 1990, the University of Colombo has carried out research on bananas, including micro-propagation through shoot-tip culture, gamma radiation-induced mutations, cell-suspension cultures and somatic embryogenesis, and ploidy analysis for detection of variation. Since 1995, investigations have been conducted to improve both Embul and Cavendish banana cultivars. Two selections of banana have been made for early fruiting and short height. Micropropagated plants of the selections were tested for trait stability until the second generation. Mass production of plants is in progress. Thus, indexing and testing of plants for viruses, i.e. BBrMV and BSV, has become essential, since virus-free indexed mother stocks are required for micropropagation. Techniques were adopted for routine testing of mother stocks, and random testing of micropropagated plants for BBrMV by DAS-ELISA with imported commercial kits.

The present study was aimed at the development of a low-cost ELISA detection kit. Anti-serum for BBrMV, produced by the

Queensland Department of Primary Industries (QDPI), Australia, was tested as the coating antibody to replace the Agdia commercial kit. Results showed a relatively high efficiency with the QDPI antibody. Work is also in progress to make an alkaline phosphatase-conjugated antibody to replace the test kit. With the production of local antiserum, it is expected that an effective low-cost local diagnostic kit could be developed for the routine indexing of banana plants for BBrMV. This would facilitate the identification of virus-free mother stocks for micropropagation. However, purification of the virus extract is still a limiting factor for obtaining the antigen.

INTRODUCTION

Importance of Bananas in Sri Lanka

Banana (*Musa* spp.) is the most widely cultivated and consumed fruit in Sri Lanka. It is also an attractive perennial fruit crop for farmers due to its high economic gains throughout the year. Even rice fields are being converted to grow banana, as it gives more economic benefits, requires relatively less water and less input than rice, and gives higher returns than rice. The farmer's net profit has gone up about four times compared with rice. In addition, not only is there water saving from banana cultivation, but also the nutritional level in farmers' families has improved due to the habit of increased fruit consumption. Also, there is a reduction in the use of chemical pesticides in banana cultivation, which has a positive impact human on health and the environment. Banana has been given the highest priority amongst Other Food Crops (OFC). Out of the 28 local cultivars, the cultivar Embul (Mysore-AAB) is in the highest demand. Kolikuttu (Silk-AAB) and Ambon (Cavendish-AAA), grown especially in the mid-wet zone of the country, are also in considerable demand.

Currently, nearly 50,000 ha of land is under banana cultivation in Sri Lanka, and the annual banana production is around 450,000 metric tonnes. Until recently, banana cultivation was limited to very small plots, but now large fields are being established. More and more rice farmers are switching to banana cultivation due to the high profit margin and smaller amount of work needed in the field. During the last six years, 2500 ha have been converted to banana cultivation .

Since 1990, under the University of Colombo research programme, research and development work has been carried out on banana micropropagation through shoot-tip culture and mass production of plants. The well developed technology has already been transferred to the farmers.

Gamma Irradiation for Improved Cultivars

Since 1995, under an IAEA/TC program, shoot tips of both Embul and Cavendish cultivars have been gamma irradiated to develop mutant cultivars such as shorter height and earlier fruiting. The irradiation of *in vitro* shoot tips with 45 Gy resulted in two selections, one with reduced height and the second with earlier fruiting. Both 'Embul' and Cavendish types started to bear fruits after six months from planting . The selected mutants were multiplied by micropropagation, and tested for genetic stability of the newly acquired traits.

Virus Diseases

In Sri Lanka, four major viruses are primarily responsible for poor banana cultivation: BBTV, BBrMV, BSV and CMV. The occurrence of BBrMV and BSV has been reported rather recently. Nevertheless both diseases are spreading widely, especially in the dry zone of the country. A recent survey by the Department of Agriculture revealed a variable pattern of occurrence of these viral diseases in different cultivars in two main banana-growing areas. For example, the BBrMV infection rate was 2.1 and 1.4%, respectively in cv. Embul and Puwalu, while both cultivars had 3.5 and 1.8% infection rate respectively. Among all the varieties evaluated cv. Embul turned out to be the most highly susceptible cultivar.

Some reports have indicated that BBTV and CMV occur prominently in relation to BBrMV and BSV. The use of infected suckers for the expansion of cultivation was the main reason for the virus spreading. Since micropropagated banana plants are not yet widely used, the chances of virus spreading through this route are negligible. However, this might change in the near future with the use of commercialised tissue-culture technology. Thus, the use of virus-free mother plants in micropropagation becomes very important for the spread of banana viruses.

Banana Bract Mosaic virus (BBrMV)

BBrMV was first reported from the Philippines in 1979 and during 1996 was also found in Sri Lanka . Thomas et al. confirmed the presence of the causal agent of the disease in the widely cultivated local variety Embul (AAB group, Mysore). In Sri Lanka, the disease is more prevalent in cultivations which are not properly managed, but the impact on the yield is significant.

Since cv. Embul is now being micropropagated, virus-indexed mother plants have become essential. At the same time, the spread of virus diseases in existing traditional small-scale plantations requires monitoring. However, a banana virus indexing kit was not available in the country until recently.

A screening method for the detection of BBrMV has not been adopted (up to 2000) in Sri Lanka, and its diagnosis depends on visible symptoms. Screening should be done with an ELISA test or a PCR-based molecular kit using specific primers. Preferably, an ELISA screening test should rely on locally developed antibodies, since this is more economical for the routine testing of mother plants, for commercial banana micropropagation, and for field cultivation as well.

METHODOLOGY

Field Testing of Selected Mutants Produced by Gamma Irradiation

Cv. Embul plants, selected for early fruiting, were proliferated *in vitro* up to M_1V_8, and 1000 regenerated plants were transferred to the fields after acclimatization. A dry zone in southern Sri Lanka

(Hambantota) and a mid-country wet zone (Ratnapura) were selected for field experiments. Ploidy analyses of plants were carried out using a Flow Cytometer (PARTEC PA II) to detect variation in ploidy level .

Screening for Bract Mosaic Virus (BBrMV)

BBrMV disease symptoms

Disease symptoms of BBrMV were identified in field-grown banana plants at different growth stages. Symptoms on the pseudostem, petioles, leaf sheath and bracts were observed (Figure 1). In the initial stages, symptoms would appear on the leaf petioles and sheath, but when severely affected would appear on the bracts. Samples, mainly from cv. Embul (AAB), were collected for laboratory testing.

Use of Agdia commercial ELISA kit

The commercial ELISA kit, Agdia® (Agdia, USA) was tested, which used DAS-ELISA with polyclonal antibodies (from rabbits) for both the coating antibody and alkaline phosphatase-conjugated antibody. Positive and negative controls were treated according to the manufacturer's instructions. Molecular biology grade chemicals (Sigma) were used as required. Microtitre plates provided with the kit were used and colour development was detected visually, as well as with an ELISA plate reader (BIORAD, 550) at 405 nm.

Several samples with BBrMV symptoms were tested, and different tissues of the same infected plant were also tested. The relationship between the colour development by ELISA and the intensity of disease symptoms was also investigated.

Testing the antibody from antiserum of the QDPI laboratory

Even though the commercial kit gave the expected results, a locally developed kit would be more economical for routine testing. Therefore, a local kit was developed with polyclonal antibodies produced by QDPI, Australia. At the initial stage, this polyclonal antibody serum was used in several dilutions to replace the coating antibody of the commercial Agdia kit, to optimise the dilution and examine the possibility of using the QDPI serum in future.

Sample preparation, extraction and purification of BBrMV

Leaf laminar tissue samples of local banana 'Embul' were collected and stored at 4°C for a maximum of 2 days, for the extraction and purification of BBrMV. All steps of sample preparation were carried out at 5°C, unless otherwise mentioned. Samples taken from the refrigerator were immediately frozen in liquid nitrogen, and were finely ground by pestle and mortar.

BBrMV extraction was done according to the procedure given by Thomas *et al.* . It involved high-speed centrifugation at 10,000 *g* (HIMAC SCP 85G) and ultracentrifugation at 125,000 *g* (HIMAC CR 21G). Equilibrium centrifugation in CsCl was done at 126,000 *g* for further purification. Gradients were collected and concentrations were measured by a UV spectrophotometer (JASCO, Model 7800) at 260 and 280 nm. The fractions with BBrMV were stored at -20°C for future evaluation of purity, and used to produce antiserum and antibodies.

RESULTS AND DISCUSSION

Field Testing of Selected Mutants induced by Gamma Irradiation

In the second generation, both early fruiting and short height traits appeared in 80% of Embul (AAB, Mysore type) plants. The other 20% flowered in the seventh or eighth month. With Cavendish, early fruiting was observed in the wet zone under good growing conditions. But in the dry zone and wet zone with poor soils, flowering was observed in the eighth month. Hence it appears that agronomic practices and soil quality have an influence on the selected cultivar. With fertile soil, and agronomic practices recommended by the Department of Agriculture , early fruiting was observed in the selected banana plants. Ploidy analysis of the selected plants did not show any variation in the position of the peaks, indicating no change in ploidy levels of cultivars with improved traits .

Early fruiting and harvesting of micropropagated banana plants saved at least one month compared with conventionally grown planting material, which usually needs 8 months to flower. Thus, the number of ratoons in two years becomes three instead of usual two, thereby increasing the income of farmers by 25%, which is equivalent to about US$ 350 per hectare per year.

Screening for BBrMV

Disease symptoms

All the currently reported disease symptoms were observed in infected plants in the field, but in varying degrees. Spindle-shaped purple or dark red coloured streak patterns on the pseudostem, in addition to dark spindle-shaped streaks on bracts, were the most common symptoms of infection.

The mature plants with inflorescences having black or reddish brown streaks on the outer surface of the open bracts also had streaks on the pseudostem. The splitting of the base of young suckers could also be due to other banana virus infections . Spindle-shaped streak patterns could also occur on the lower side of the petiole of the leaf, and raised veins of the upper surface of the leaf lamina is another symptom. The upper leaf surface may also show chlorotic spindle-shaped yellowish streaks, but it was difficult to relate it to BBrMV infection only as other virus infections may produce similar streaks.

Indexing for BBrMV using the Agdia kit

The Agdia kit gave positive results for all the infected samples. It was noted that the colour intensity of the test results increased with the increase of the severity of disease symptoms. Even in the same plant, some tissues without disease symptoms gave negative results, while parts with symptoms gave positive results. For example, the bracts with streaks gave positive results with the Agdia kit, but flower parts under the same bract were negative. However, for general diagnosis purposes, the commercial Agdia kit could be recommended.

Routine testing of mother stock and random testing of micropropagated plants for the BBrMV by DAS-ELISA is now being carried out. Thus, it is now possible to use BBrMV-free mother stocks for mass production of plants through shoot-tip culture. Development of a local ELISA kit is in progress, and the establishment of the virus-indexing procedure would greatly assure the sustainability of the tissue-culture program, which would also use the mutants developed through mutation breeding.

Use of QDPI antibody

The available polyclonal QDPI antibody at a dilution of 1:400 without further purification was successfully used to replace the original coating antibody of the Agdia commercial kit.

The dilution rate of the coating antibody of the commercial kit was 1:200. Therefore, the QDPI polyclonal antibodies would be more economical if other major components of the kit, e.g. alkaline phosphatase-conjugated antibody, could also be prepared from it. The next step would be the preparation of the enzyme conjugate using QDPI antibodies after purification.

Virus isolation and purification

According to Thomas et al. , the virus-containing fraction should give a ratio of 1.17, and therefore, only fraction 3 may contain the virus.

However, further evidence is required to confirm that fraction 3 is free from other plant or tissue contaminants. If banana plant contaminants were present, antisera against them would be produced and ELISA would give false positive results. Currently, the virus purification step has become the limiting factor for pure antigen extraction for the antibody production process. The absorption ratio of CsCl extracted fractions is given in Table below.

Table: The absorption ratio of different CsCl extracted fractions

Fraction number	1	2	3	4	5	6
Absorption 260 nm:280 nm	1.16	1.19	1.17[a]	1.18	1.10	1.20

[a] virus containing fraction according to Thomas et al.

CONCLUSIONS

In Sri Lanka, radiation-induced mutations have been used for banana improvement. A mutant, developed from the cv. Embul (AAB, Mysore type), has been isolated with traits including shorter height and early fruiting. The latter allows the reduction of the fruiting period by one month and thus enables farmers to have four harvests in 24 months from three ratoons. An increase of one ratoon crop has raised annual income of banana growers by about 25%. As a result of the introduction of tissue culture-derived banana plants, the selected mutants were readily multiplied and made available to the growers. There was an increase in demand for micropropagated banana plants, which had a significant impact on the local economy. The most important problem encountered with the production of tissue cultured banana plants during the last few years was the selection of virus-free mother plants.

This project has developed the basic infrastructure at the University of Colombo to develop ELISA diagnostic kits for the detection of BBrMV through technology transfer. The results obtained so far strongly support the potential of developing a local testing kit for detection of BBrMV in Sri Lanka. Experiments are in progress on the extraction and purification of the BBrMV from local banana cultivars.

FIELD EVALUATION OF POTENTIAL MUTANTS OBTAINED AFTER GAMMA IRRADIATION

Bananas and plantains provide an important carbohydrate source in the Cuban diet. Low yields and susceptibility to diseases (mainly black Sigatoka, caused by the fungus *Mycosphaerella fijiensis*) of current cultivars render the development of new varieties a high priority. Biotechnological and nuclear techniques have been successfully applied on shoot tips, resulting in the generation of improved cultivars. Potential induced mutants derived after gamma irradiation were tested in the field.

A dwarf-like mutant of the cultivar 'Parecido al Rey' (clone V6-44) was obtained. Some mutants (Parecido al Rey variant 6-32 and 6-44 and Gran Enano variant 3-2) were selected for further field evaluation due to their tolerance to black Sigatoka disease, as well as their higher yield compared to non-irradiated control plants. These characters appeared to be unstable during following field cycles. To reduce the frequency of chimeras obtained after irradiation and to obtain a higher multiplication rate of irradiated samples, cell suspensions were established and used as starting material for mutation induction. The regeneration capacity of the established 'Navolean' cell culture was quantified. Homogenized samples of 500 µl cell suspension gave rise to between 1300 and 2650 embryos with a germination frequency of 20.7%. Percentages of survival *ex vitro* amounted to 95%. Field performance and genetic stability of plants regenerated from irradiated cell suspensions are currently under investigation.

Bananas and plantains are among the most important food crops throughout the world, and constitute an important carbohydrate source in the Cuban diet. However, low yields and susceptibility to diseases of local cultivars (mainly black Sigatoka caused by the fungus *Mycosphaerella fijiensis*) necessitates the generation of improved clones. Due to infection by black Sigatoka and the lack of resources to control this disease, the banana and plantain area in Cuba has been reduced, and banana cultivars replaced by black Sigatoka-resistant FHIA hybrids and cooking bananas (ABB group) despite the high preference for almost all plantain (AAB group) cultivars. Hence there is an urgent need to include plantains in the Cuban banana and plantain improvement program .

The genetic improvement of banana and plantain through conventional breeding has been hampered by the lack of a profound understanding of the genetics of these plants and their pathogens, the interaction and co-evolution of host and pathogens, and their high sterility, etc. . Biotechnological and nuclear techniques have proved their merit for the induction of genetic variability and mutant selection. Gamma irradiation of *in vitro* shoot-tip cultures can induce changes in one or a few characteristics of the treated explants without altering their unique cultivar traits . Irradiation of cell cultures followed by regeneration through somatic embryogenesis opens great possibilities, not only in relation to a high multiplication rate but also in the context of avoiding chimerism due to the presumed single cell origin of suspension-derived plants .

Table: Medium composition for* in vitro *culture initiation, multiplication (used at IAEA or INIVIT) and rooting

	Culture initiation	Multiplication IAEA[a]	Multiplication INIVIT[b]	Rooting
Inorganic salts and vit MS	4.4 g/l	4.4 g/l	4.4 g/l	
Half strength inorganic salts				2.2 g/l
Thiamine	1 mg/l	1 mg/l	1 mg/l	
Vit MS				0.1 g/l
Sucrose	40 g/l	40 g/l	30 g/l	20 g/l
Indole 3-acetic acid (IAA)	0.88 mg/l		(0.65 mg/l)	
N6-benzylaminopurine (BAP)	2.25 mg/l	4.5 mg/l	4 mg/l	
Indole 3-butyric acid (IBA)				0.2 mg/l
Gelrite	2 g/l	2 g/l		
Agar			7 g/l	8 g/l

[a] Multiplication medium used for 'SH 3436-L9'and 'Paracido al Rey'

[b] Multiplication medium used for 'Gran Enano' and 'Burro CEMSA'

The present work was performed to obtain mutants characterized by reduced height, increased yield, and higher resistance or tolerance to black Sigatoka than the original clone. Furthermore, plant regeneration via somatic embryogenesis was optimised for its application in genetic improvement programs.

MATERIAL AND METHODS

In Vitro Establishment and Gamma Irradiation

Shoot-tips as starting material

Plant material selected for introduction *in vitro* consisted of vigorous and healthy suckers of commonly grown cultivars in Cuba. *In vitro* cultures of 'SH 3436-L9' (somaclonal variant of FHIA hybrid 'SH 3436', AAAA group), 'Parecido al Rey' (AAA group), 'Gran Enano' (AAA group) and 'Burro CEMSA' (ABB group) were established by placing their apical meristems onto MS-based culture initiation medium.

Once successfully initiated, explants were cultured on multiplication medium until the required amount of starting material for irradiation was obtained. Shoot tips consisting of the apical dome and 2-4 leaf primordia were excised from rooted plants or multiple shoot-tip cultures and submitted to gamma irradiation in Cuba ('SH 3436-L9', 'Gran Enano' and 'Burro CEMSA') or at IAEA ('Parecido al Rey').

To determine the optimal irradiation dose for 'SH 3436-L9', a dose-response curve was generated by irradiating 40 shoot-tips to doses of 0, 25, 35, 45 and 60 Gy respectively. Treated explants were then cultured on multiplication medium for 21 days, and the surviving shoot-tips and the number of proliferating buds counted. From these data, the mean lethal doses (LD_{50}) for irradiation of 'SH 3436-L9' shoot-tips could be determined by Probit analysis .

The optimal irradiation doses to induce mutants of 'Paracido al Rey' (60 Gy), 'Gran Enano' (35 Gy) and 'Burro CEMSA' (35 Gy) had already been obtained from earlier experiments. After mutagenic treatment, shoot-tips were cultured for three cycles on fresh multiplication medium. The M_1V_4 multiple shoot cultures obtained were separated into individual shoots and transferred onto rooting medium for 2-3 weeks. Rooted plants were hardened in the greenhouse for approximately 60 days.

If the desired number of plants for field evaluation (>1000) was not reached, plants were multiplied by corm sectioning so that each section contained at least one bud. This material was incubated in chambers under high humidity conditions. This technique is called SPC (Seed Production Chamber) and was developed at INIVIT. For each explant, 45 days of culture resulted in 10-15 additional 'seeds' for field evaluation.

Table: Medium composition for culture initiation and multiplication of meristem cultures, embryogenesis induction and maintenance (ZZ), regeneration (RD1) and germination of somatic embryos and rooting of embryogenic cell suspension derived plants

	Culture initiation	Multiplication	ZZ semi-solid	RD1	Germination	Rooting
Inorganic salts and vit MS	4.4 g/l	4.4 g/l			4.4 g/l	4.44 g/l
Half strength of inorganic salts			2.2 g/l	2.2 g/l		
Thiamine	1 mg/l					
Myoinositol				100 mg/l	100 mg/l	100 mg/l
Sucrose	30 g/l	30 g/l	30 g/l	30 g/l	30 g/l	20 g/l
IAA	0.9 mg/l	0.2 mg/l			1 mg/l	
BAP	1.1 mg/l	23 mg/l			(1) 0.3 mg/l (2) 0.4 mg/l (3) 0.5 mg/l (4) 0.6 mg/l	
2,4-D			1.1 mg/l			
Zeatin			0.2 mg/l			
Ascorbic acid		10 mg/l	10 mg/l	10 mg/l		
Gelrite		2 g/l	2 g/l	2 g/l	2 g/l	2 g/l
Agar	6 g/l					

Embryogenic cell suspensions as starting material

Scalp-derived embryogenic cell suspension of 'Navolean' (AAB group) were developed according to Schoofs . This methodology requires a material preparation phase followed by induction of embryogenesis, and initiation and maintenance of the embryogenic cell suspension. After inoculation of apical meristems onto culture initiation medium, *in vitro* meristem cultures were optimised by subsequent cycles on BAP-rich multiplication medium.

Optimal explants ('scalps') for induction of embryogenesis were excised from the uppermost part of the meristem culture and cultured on semi-solid ZZ medium. Embryogenic calli formed on embryogenesis-induced explants were transferred to liquid ZZ medium (semi-solid ZZ medium without gelling agent) in Erlenmeyer flasks with a volume of either 10 or 25 ml. If the quantity of embryogenic material was so low that the minimal inoculum density required for a successful establishment of an embryogenic cell suspension was not reached, two or three embryogenic calli (derived from different induced explants) were cultured together in one Erlenmeyer flask of 25 ml. Liquid cultures were kept on a rotary shaker (70 r.p.m.) under a light regime of 16:8 light:dark cycle. During the early initiation phase of the embryogenic cell suspension, the culture medium was refreshed every third day. Four weeks after initiation, renewal of liquid ZZ medium was done at bi-weekly intervals. As soon as the cell cultures reached the phase of mass multiplication, the settled cell volume (SCV) was adjusted to 3% at the onset of the subculture period.

To determine the regeneration capacity of the established cell cultures, a homogenized sample of 500 μl suspension (3% SCV) was plated onto RD1. The weight of a representative sample of a 45-day-old regenerated cell culture was measured, and the embryos counted under a stereomicroscope. In this way, the relationship between the weight and the number of embryos could be determined. After embryo differentiation, 12 samples each containing 0.1 g globular embryos were incubated on four different but similar germination media. Percentages of embryo germination were recorded 30 days later. Once explants with leafs and/or small roots were formed, shoots were transferred to rooting medium. Twenty-day-old rooted plantlets that had reached a length of 5 cm were grown *ex vitro*, and surviving percentages were evaluated after 45 days.

Before irradiation, the SCV of the Navolean embryogenic suspension was increased from 3 to 25%. Samples of 500 µl of this concentrated suspension were irradiated at 10, 20 and 30 Gy respectively. After irradiation, the SCV of the suspension was again adjusted to 3%.

At present, efforts are concentrated on the establishment of embryogenic cell suspensions and plant regeneration via somatic embryogenesis for cultivars varying in their resistance to Fusarium and Sigatoka diseases (Zanzibar, CEMSA 3/4, Burro CEMSA, FHIA 21, Calcutta 4, Niyarma Yik, Pisang Berlin, Pisang Lilin, Rose, Bluggoe, Manzano, Gros Michel, Pisang Ceylan, Pisang Rastali, Yangambi Km5, Pisang Tugi and Pisang Mas). Cell cultures are developed according to the scalp methodology and/or the male flower technique .

Field Evaluation of Potential Mutants Obtained by Gamma Irradiation

After hardening in the greenhouse, control plants (non-irradiated) and irradiated plants derived from shoot tips (at a dose corresponding to LD_{50}) were grown in the field. At flowering and harvest each of the cultivars involved (SH 3436-L9, Parecido al Rey, Gran Enano and Burro CEMSA) was evaluated for plant height (cm), pseudostem circumference (cm), weeks until flowering, number of hands per bunch, number of fingers per bunch, bunch weight (kg) and incidence of black Sigatoka disease at flowering, according to internationally accepted criteria . Statistical analysis of the mean values obtained was done by Duncan's Test. Based on desirable agronomic traits such as low height, early flowering, high yield and/

or increased tolerance towards black Sigatoka, potential improved variants were selected for further field evaluation. Prior to the following field cycle, selected clones were multiplied by SPC until the required number of hardened plants was obtained. The genetic stability of potential variants was deduced from their behaviour during subsequent field cycles. Similarly, plants derived from shoot tips, from suspension culture, and control plants of Navolean were evaluated for their field performance six months after planting.

RESULTS AND DISCUSSION

Shoot-tip Derived Plants as Starting Material

Increasing irradiation of shoot tips of SH 3436-L9 resulted in fewer surviving explants. Simultaneously, a linear decrease in the proliferation rate of treated explants was noted (112,68,49,29 and 4 buds counted on explants irradiated at 0,25,35,45 and 60 Gy respectively).

Figure: Dose-response curve of 'SH 3436-L9' shoot-tips established 21 days after exposure of explants to different irradiation doses (40 shoot tips were treated per irradiation dose).

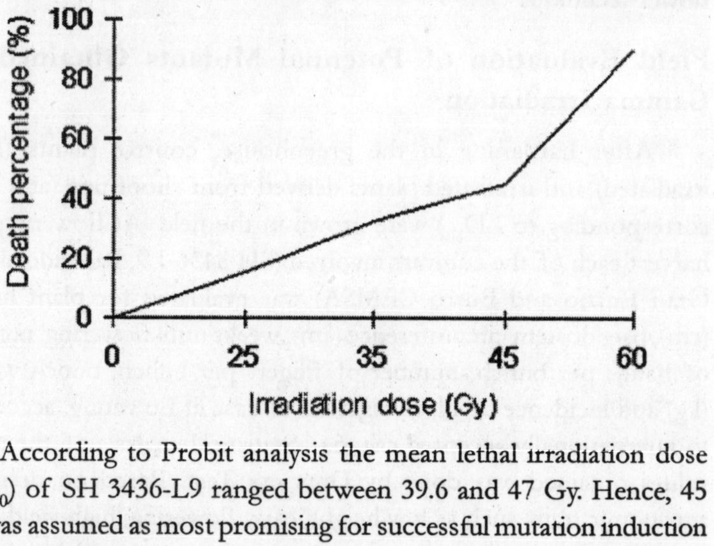

According to Probit analysis the mean lethal irradiation dose (LD_{50}) of SH 3436-L9 ranged between 39.6 and 47 Gy. Hence, 45 Gy was assumed as most promising for successful mutation induction of SH 3436-L9 shoot tips. Consequently, only plants derived from explants submitted to this irradiation dose (45 Gy) were evaluated in the field.

Results on mean field performance values of potentially induced mutants and control plants of SH 3436-L9, Parecido al Rey, Gran Enano and Burro CEMSA are summarized in Table 3. In a first harvesting cycle, nine irradiated clones of SH 3436-L9 were selected for further evaluation due to their lower height (280-300 cm compared to a mean height of 306 cm for the control), earlier flowering (41-42 weeks compared to a mean value of 43.2 weeks for the control), number of hands per bunch (11-13 compared to a mean number of 11 for the control), number of fingers per bunch (177-225 compared with a mean number of 170 for the control) and bunch weight (32.8-36.2 kg compared to 32 kg for the control). However, these advantageous agronomic traits appeared to be unstable during the second field cycle (data not shown).

For Parecido al Rey, a total of 900 irradiated plants and a similar number of control plants were evaluated for their field performance. In the first cycle, 58% of potentially induced mutants showed phenotypic variations. In particular, bunches were deformed (about 50%). Dwarf and extra-dwarf plants were found, as well as deformed leaves and split fingers. Probably irradiation doses were too high, as each genotype has its optimal LD_{50}. Variants could also be associated with chimerism and the number of subculture cycles on proliferation medium . Because of this very high frequency of variation, no selection was carried out in the first cycle, but material was propagated for further evaluation. In the second cycle eight irradiated clones of Parecido al Rey seemed to perform better in height and yield than the control plants. Two of these better performing clones (V6-32 and V6-44) showed also some tolerance to black Sigatoka (Table 3) and were thus evaluated in a third cycle. In the third evaluation cycle, clones V6-32 and V6-44 were significantly smaller than the control (176.5 and 160.1 cm versus 214.2 cm) and flowered earlier (40.6 and 40.1 weeks versus 42.1 weeks for the control) (Table 3). Since, in addition, a significantly higher mean bunch weight was obtained for variant V6-44 (23.8 kg compared to

20.0 kg for the control), this clone was further evaluated for its field performance. In a fourth cycle, the yield of mutants of V6-44 was substantially higher but not different from the control (data not shown). Among the initial population of 3000 Gran Enano plants evaluated for field performance, two mutants (V2-9 and V3-2) were selected because of their shorter height (data only shown for variant V3-2). These mutants were multiplied by SPC until 70-80 plants per mutant were obtained. In the second cycle, mutant V3-2 remained significantly smaller than its control (184 cm versus 192 cm) and seemed to be less susceptible to black Sigatoka than its control. Other agronomic characteristics remained unaffected (Table 3). These two important characteristics were, however, unstable in the third cycle. A total of 1730 (irradiated and non-irradiated plants) of Burro CEMSA were grown in the field. Based on desirable agronomic traits such as early flowering and lower plant height, two lines (V2-16-1 and V4-20-2) were better performing during a first field evaluation cycle. Selected lines were multiplied via SPC and followed for their field performance during a second growing season. During the second field cycle, these two selections lost their superiority over the control. All these results obtained in different cultivars indicate that selected materials were unstable during subsequent field cycles.

Establishment of Embryogenic Cell Suspensions and Field Evaluation of Gamma-irradiated Suspension-derived Plants

Optimised *in vitro* meristem cultures of Navolean were obtained after three subsequent cycles on BAP-rich multiplication medium. From these highly proliferating meristem cultures, a total of 150 scalps were excised and induced for embryogenesis. One week after induction of embryogenesis, explants showed no evident changes. However, 4-5 weeks after placement of the explants on ZZ medium, the original explant had doubled in size. From the fourth to the sixth week, yellowish-white nodular tissue appeared at the surface of the induced culture. White meristematic globules (nodular callus) gave rise to the first somatic embryos and the formation of embryogenic complexes similar to those previously described. In seven out of 150 cases, embryogenic calli suitable for transfer to liquid medium were observed. The embryogenic frequency obtained (5%) is considered to be high for bananas and plantains, since these crops are rather recalcitrant towards embryogenesis. Transfer of embryogenic calli to either 10 or 25 ml Erlenmeyer flasks revealed a higher proliferation rate of embryogenic cell clusters and subsequently more rapid establishment of the cell culture if the smallest container was used for initiation. Since (a) the quantity of available embryogenic cell material is often low, and (b) a minimal inoculum density is required for successful initiation in liquid medium, the use of 10 ml Erlenmeyer flasks is an important and advanced step in the establishment of embryogenic cell suspensions. Cell cultures which had reached the phase of mass multiplication consisted of a great number of actively dividing spherical cells and heterogeneous translucent and non-translucent aggregates similar to those previously described [8,20]. Forty-five days after transfer of suspension samples (500 µl, 3% SCV) onto regeneration medium RD1, 0.1 g embryogenic tissue contained between 325 and 663 globular embryos. Transfer of regenerated cultures (0.1 g per sample) onto germination medium 1 (containing 0.3 mg/l BAP and 1.0 mg/l IAA as plant growth regulators) was found to be significantly better (Duncan's test, *P* £ 0.05) than the other germination media tested.

Table: Mean number of embryos counted 30 days after transfer of 0.1 g of embryogenic tissue onto germination media differing in BAP concentration

Germination medium	Mean number of embryos
1	77.6a
2	74.1b
3	52.6c
4	28.4d

Number of repetitions is 12

Statistics according to Duncan's test (*P* £ 0.05).

The use of germination medium 1 resulted in a germination frequency of 21% and is similar to reports on the diploid cultivar *Musa acuminata* (20-36%) but higher than for the clone 'French Sombre' (10-14%) . An increase in BAP concentration from 0.3 to 0.6 mg/l resulted in a decrease of embryo germination from 21 to 7.6% for the cultivar Navolean (AAB group). Six weeks after hardening *ex vitro* 95% of the plants had survived.

Compared to the control, a growth increase was remarked for Navolean suspensions irradiated with a dose of 10 Gy. Since the suspension irradiated at 30 Gy was still regenerating, it was not possible to determine the mean lethal and subsequent optimal irradiation dose for Navolean cell cultures. Field evaluation of suspension-derived irradiated (0-30 Gy) and control plants are currently under investigation. Besides the highly regenerable Navolean cell suspension, cell cultures were also successfully established for Zanzibar and CEMSA3/4. The development of embryogenic cell suspensions for other cultivars is in progress.

15

GENETIC MECHANISM, FREQUENCY AND APPLICATION

Clonal selection of banana (*Musa acuminata AAA*, cv. 'Grande Naine') was performed under the marginal climatic conditions for banana cultivation in the Western Galilee region of Israel. The initial selection was carried out on 300 mats originating from six mother clones multiplied by meristem culture. Five clones were selected for further experiments. In the third growth cyclè the total number of bunches per hectare of three out of the five selections, exceeded that of the control, while the remaining two selections performed similar or below the control. In the fourth cycle all the selections out-performed the control. Average bunch weights of two of the selected clones were significantly higher than the control. The total yield, derived from bunch weight and the number of bunches per mat of each of the selections, was higher than that of the control in both the third and the fourth cycle.

Hundred and fifty hectares of micropropagated plants derived from the selected clones were planted in the Philippines, and their performance was evaluated and compared to that of a local selection of 'Grande Naine'. The yield of the Israeli selected clones in the Philippines was significantly higher by an average value of 18 percent compared to the local clones.

Using degenerate primers, a fragment of a retro-transposable element (BR-1) that was induced by extensive in vitro culture was detected in *Musa.* Transcriptional activation of the banana retro-transposon was detectable by RT-PCR in the 'Long and Narrow Leaf ' mutant (LNL). The frequency of occurrence of BR-1 was analyzed in genomic DNA of the LNL mutant and in its originator non-mutant mother plant. The analysis revealed amplification of BR-1 in the mutant compared to the mother stock DNA. We postulate that retro-transposable elements are at least in part responsible for the occurrence of the LNL mutant.

INTRODUCTION

Cultivated bananas are natural selections originating from the center of origin of the genus *Musa* in the Asian Pacific region. The edible types comprise a range of natural hybrids originating from the two species *Musa acuminata* (A genome) and *Musa balbisiana* (B genome). Most cultivars are triploids (AAA, AAB or ABB genomes), parthenocarpic and sterile, though in the international trade of dessert bananas, the vast majority belongs to the AAA 'Cavendish' subgroup. Despite the intricacy of breeding infertile plants, a number of new banana cultivars have been released over the last three decades. All the new hybrids that have been accepted by growers comprise a mixture of A and B genomes. While the use of the new hybrids is rising, consumer acceptance is limited to local (not export) markets.

The main focus of banana breeding programs for the export industry is resistance to diseases and improvements of fruit quality. The extreme susceptibility of the old cultivar 'Gros Michel' to *Fusarium* wilt (Panama Disease) forced breeders and producers of export sectors of the industry to shift to a more resistant variety. The newly created cultivars should confer resistance or tolerance to the major diseases, in order to reduce the costs and extensive use of fungicides and nematicides. Moreover, eruption of new races of pathogens can endanger the existence of the entire banana industry, due to the narrow genetic pool of selected clones. The outbreak of viral diseases constitutes another threat to the banana industry in most production areas.

Introduction of *in vitro* culture allowed production of pathogen free plants and widened the genetic variability of existing cultivars, as a result of somaclonal variation. At the same time, due to the variation that occurs during the *in vitro* propagation, plants of the same cultivar selected in different locations often differ in some characteristics. Thus, the *in vitro* technique allows local genetic selections. An example of a banana cultivar improved by somaclonal variation was demonstrated in Taiwan by who selected somaclones resistant to *F. oxysporum*. These clones were further improved by somaclonal variation for better field performance.

The high incidence of "off-types" produced by banana meristem culture is a major concern to commercial growers. At the same time, somaclonal variation is an important tool for the improvement of banana germplasm. Banana plant propagators and researchers have learned, over the years, to reduce the frequency of "off-types" to a manageable level. The majority of the banana "off-types" in the Cavendish subgroup appear as 'dwarf' or 'giant'. It has been demonstrated that both are associated with sensitivity to gibberellic acid . Most types of the in-vitro generated mutations in Cavendish result from a long duration in tissue culture and from high number of cycles in the multiplication phase of the meristem culture. The specific genes responsible for these mutants have not been identified so far. Several gibberellin insensitivity mutants have been isolated and characterized in other plant species, but none of these have been associated with somaclonal variation. Unlike Cavendish types, an array of phenotypic variations was reported in plantains. Most variations in plantains were related to the morphology of the inflorescence and fruit . Thus, it seems that the highly frequent types of variation are genotype specific.

The Western Galilee in northern Israel is on the border line of commercial banana production. In the summer months the solar radiation reaches very high levels, the ambient temperatures are high and rainfall is practically nil. During winter nights the temperatures occasionally drop below the freezing point. Under these low temperatures banana plants are severely damaged, but nevertheless production remains adequate. Sucker selection in Israel is practiced during the summer months and is aimed to restrict flowering to mid summer of the next year and harvest to the following winter months. Sucker selection is performed by experienced growers and is based on the number of accumulated leaves on specific dates. The "selector" determines the number of suckers to be left for the next cycle in each mat, according to physiological conditions of the plants in the specific mat and its relations with adjoining mats. Yield is determined mainly by the number of bunches/ mat (usually 1-3 per mat) and the weight of each bunch.

In tropical regions like the Philippines, the number of bunches per mat is fixed (usually only one). Therefore clonal evaluation is influenced by bunch weight only.

Due to the relatively "harsh" climatic conditions and unique topographic location combined with strict quarantine measures, the Western Galilee is currently free of the major banana pests and diseases, including Black and Yellow Sigatoka, Banana Bunchy Top Virus, Banana Streak Virus, Panama disease and the most damaging nematode *Radopholus similis*. The vectors that transmit these diseases seem unable to overcome neither the cold winter temperatures nor the low humidity in the long and dry summers.

In spite of the work reported thus far on clonal variation in banana, the genetic aspects of this phenomenon are unknown. The lack of understanding of the genetic events leading to somaclonal variation puts severe constraints on producers of tissue cultured banana plants. At the same time it opens an opportunity for selection of elite clones. In this investigation we attempted to elucidate the genetic mechanism of clonal variation that is induced by tissue culture.

MATERIALS AND METHODS

Initial Evaluation

The initial selection was performed on 300 mats of **'Grande Naine'** produced from six mother clones, propagated by **meristem** culture. The *in vitro* cultures were initiated from apical meristem explants and were propagated and rooted according . Hardening, potting and growth prior to transfer to the field were carried out as practiced commercially by Rahan Meristem. After three months of *ex vitro* hardening, the plants were planted in the Western Galilee Banana Experimental Station.

Mats were arranged in wide "tram-line" paired rows . The distance between each pair was 6.0 m and between rows within the pair 3.0 m. The planting distances within the row were 2.7 m. Thus, the total density was 832 mats/ha. Drip irrigation, fertilization and all other agrotechnical procedures were according to stāndard methods practiced in the Western Galilee.

The evaluation was based on data collected from each individual mat over a six year period. The criterion for selection was yield per mat, calculated by adding the weight of the bunches produced in the mat. Four clones, '5-1', '6-6', '37-5' and '42-5', which performed exceptionally well, were selected for further evaluation. A fifth clone '17-1' was selected separately in response to cold resistance characteristics (data not shown).

Clonal Evaluation

Based on the results of the above experiment, the best clones were multiplied by meristem culture as mentioned above.

Seventy plants of each of the selected clones and of a control representing the average value for all 300 original mats were used for the experiment.

The planting material consisted of hardened micropropagated plants, three months after hardening. Planting distances were as mentioned above (823 mats /hectare). The plants were arranged in randomized blocks of five replicates. The numbers of bunches/ mat and bunch weights were recorded for 10 mats in each replicate (excluding the mats at the end of the rows). Altogether 20-30 bunches per replicate and over 120 bunches per clone were evaluated each year.

In the winter following planting temperatures were unusually low, hindering plant growth for two consecutive cycles. The yield in the first two years was unusually low and as a consequence the results of these cycles were ignored. The data presented in this report include the results of the third and the fourth years after planting.

Approximately 30 hectares of micropropagated plants of the selected clones were planted in a commercial banana plantation in the Western Galilee region. These 30 hectares served as a source of explants for the evaluation experiments in the Philippines. The parameters for choosing the explants within the plant population were flowering time and yield.

In the Philippines the performance of the Israeli selected clones was compared to the leading 'Grande Naine' selection, which was introduced to the Philippines from Ecuador (Ballesteros, personal communication). The local plants were also propagated by meristem culture and potted as mentioned above.

Planting scheme in the Philippines was in "hedgerows" 4.5m apart. The plants were arranged within the row in a triangular configuration at distances of 1.25m. Total density was 1973 plants per hectare.

The plants were drip irrigated and fertilized according to standard methods practiced in the Philippines. The experiment was conducted in 3 separate farms, each containing both Israeli and Ecuadorian selections at a ratio of 1:1, in 50 hectares plots. The values presented are the calculated means of all three plots.

Characterization of 'off-types'

Induction of variation by extensive duration of meristems in in-vitro conditions

Meristem culture was performed on a single 'Grande Naine' clone employing a standard protocol used at Rahan Meristem. The meristems were sub-cultured extensively (23 cycles) to induce somaclonal variations. The in vitro plantlets were transferred to a stage III medium containing 10 ppm of GA_3. Analysis of GA sensitivity was examined by measuring elongation and the distance between internodes of the *in vitro* plants following a four weeks culture period on the GA enriched medium .

Further analysis of somaclonal variation was carried out in *ex-vitro* conditions. Following acclimatization the plants were planted in 5 L pots and placed in a poly-ethylene covered greenhouse for approximately 16 weeks prior to selection of 'off-types'. Candidate plants showing distinct somaclonal variations were selected and used for further analysis. Stability of the different mutations was monitored following the growth of the selected clones in a commercial plantation in Rosh Hanikra, Israel.

DNA isolation and Southern blot analysis

Samples (2.5 g) of fully expanded leaf blade tissue were harvested and ground by mortar and pestle under liquid nitrogen. The samples were homogenized in 25 mL extraction buffer containing 4% (W/V) CTAB, 10 mM Tris-HCl pH 8, 1.4 M NaCl and 20 mM EDTA. The extracts were placed at 65^0C for 30 min. After cooling to room temperature an equal volume of chloroform:isoamyl alcohol (20:1) was added and after 15 min. of incubation at room temperature the mixture was centrifuged at 5000 rpm for 5 min. The supernatant was filtered through 5 layers of cheesecloth and an equal volume of ice cold iso-propanol was added to the filtrate. Following addition of NaCl to a final concentration of 0.1 M the samples were kept at -20^0C for one hr. and subsequently centrifuged for 15 min. at 11,000 rpm at 4^0C. The resulting pellet was resuspended in 3 mL of 70% ethyl alcohol, the mixture was centrifuged as above, and the resulting pellet was resuspended in 0.5 mL distilled water. Aliquots of ten μg DNA were digested with *Eco*RI and separated on a 1.2% agarose gel, stained with ethidium bromide and blotted onto a Nytran membrane. Transfer of DNA and hybridization was according to. The membranes were probed with a 344 bp PCR product previously isolated from banana genomic DNA using *Tos*-1 primers.

RNA isolation and RT PCR

Five grams of young leaf tissue were ground in liquid nitrogen and homogenized with extraction buffer containing 0.2 M Tris-HCl pH 8.5, 0.33 M $LiCl_2$, 10 mM EDTA and 1% PVPP. The homogenate was centrifuged for 20 min. at 7000 rpm. Following centrifugation the supernatant was filtered through five layers of cheesecloth and subsequently cold ethanol to a volume of 10% (V/V) and 3% (V/V) 3.3 M NaAc was added.

The mixture was centrifuged at 10000 rpm for 10 min. and the supernatant collected and extracted twice with an equal volume of phenol: chloroform and once in chloroform. The aqueous phase was precipitated over night at -20^{0}C in an equal volume of isopropanol and 1/10 volumes of 3.3 M NaAc (pH 6.1). Following centrifugation at 7000 rpm at 4^{0}C for 20 min. the pellet was resuspended in cold 80% ethyl alcohol, centrifuged as above and finally rinsed in absolute alcohol. The pellet was resuspended in DEPC treated distilled water to a concentration of 2 μg per μL.

The RT reaction mixture (total volume 25 μL) contained 0.5 μg RNA, 1 mM (each) of a nucleotide mixture, 40 pmoles of oligo T_{18} primer, 50 units of RNA's inhibitor, 20 units of RT enzyme (AMV - Boehringer Mannheim) and 1 μL of enzyme reaction buffer. The reaction mixture was incubated for 10 min. at 25^{0}C, 60 min. at 37^{0}C and 5 min. at 95^{0}C.

PCR was performed in a 50 μL reaction mixture containing: a 1 μL aliquot from the RT reaction, 0.4 mM (each) nucleotide mixture, *Tos*- 1 upper and lower primers (50 pmoles each) and 2.5 units of *Taq*-polymerase (KlenTaq).

The reaction was performed at 95^{0}C dissociation for 2 min. followed by 25 cycles of 30 sec. at 95^{0}C, 60 sec of annealing at 60^{0}C, 1 min. of primer extension at 72^{0}C and finally the mixture was kept at 72^{0}C for an additional 7 min. Following the PCR the samples were separated on a 1.2% agarose gel, stained with ethidium bromide and visualized with a UV lamp.

RESULTS

Initial Selection

Out of the 300 mats that were planted in 1984, five clones performed substantially better than the control (combined value of all the clones used for the selection). The yield of each of the selected clones exceeded the control.

Table: Average yield of 5 selected clones in the 6 years of the experiment, as compared to control. Yield calculated according to a density of 832 mats/ha

Parameter	Clone					
	6 - 6	5 -1	17- 1	42- 5	37- 5	Control
Bunch wt. (kg)	37.47	37.82	33.44	37.44	35.68	30.47
Bunches/ mat/ year	2.83	2.83	2.33	2.83	3.00	1.74
Calculated yield (t/ha/yr)	87.38	88.19	64.22	87.31	88.10	43.59

Five selections '5-1', '6-6', '37-5', '42-5' and '17-1' generated 35-75 % more bunches per mat compared to the average (control). The average bunch weight of all five selections exceeded the control by 2.9-7.3 kg (9.5-24%). As a consequence, the yield of the four selections was approximately twice as high as the control, while the yield of selection '17-1' was only 50% higher.

Evaluation of Selected Clones under Israeli Conditions

The total number of bunches/ha of clone '17-1', '42-5', and '37-5' in the third year of production exceeded the control ('42-5' not statistical significant), while clones '6-6' and '5-1' were below or similar to the control. In the fourth cycle (Figure 1b), all the selections performed better than the control, though only in clone '17-1' the difference in the number of bunches was statistically significant (P=0.05). Clone '6-6' generated the lowest number of bunches in the third cycle (1742 compared to 1753 in the control), but at the fourth cycle it out-performed the control by approximately 10%. Selection '37-5' generated the highest number of bunches per hectare in both growth cycles (more than 11% higher than the control).

Average bunch weights of selections '5-1' and '17-1' were significantly (P= 0.05) above the control value in both the third and fourth cycles. On the other hand, clone '42-5' generated the least weight per bunch in both cycles.

The total calculated yield, derived from the number of bunches and bunch weight, was higher in each of the clones compared to the control. The ratio between the selections followed a similar pattern.

Comparison of Selected Clones with a Local Selection in the Philippines

Although the values for bunch weight and box/stem ratio (Figure 4,5) for Israeli selections were consistently higher throughout the 20 months of the experimental period, all selections followed a similar pattern of temporal changes in fruit production.

Bunch weight of both selections gradually rose from January 1995 and reached maxima in October (Figure 4). The weight fell during 1996 in both selections. A wide and significant difference in favor of the Israeli selection appeared from April 1995 and on.

The difference reached a peak of over 30% (8.5 kg/bunch) and moderately declined in the subsequent months to approximately 5 kg/bunch (20%).

Box to stem ratio, which reflects both quality and yield, was higher for the Israeli selection compared to the local one. The difference gradually increased from January to October 1995 and slowly declined until very small differences were noted in the middle of the second year of measurements. The highest value of 2.26 and 1.8 boxes/ stem respectively for the Israeli and local selection was reached in October 1995.

Characterization of Common Somaclonal Variants in Mass Propagated Banana Plants

Somaclonal variants with both high and low sensitivity to GA were generated as a consequence of an extensive duration in tissue culture. Both 'off-types' were detectable by a relatively simple bioassay developed in our laboratory. GA sensitivity differed significantly between the different plants. The dwarf genotype had approximately 40% shorter internodes in comparison to the normal phenotype, but showed sensitivity to GA. On the other hand the 'Extra Dwarf' phenotype was insensitive to the presence of GA in the medium.

Table: Influence of GA_3 on "internode" length (cm) of normal in-vitro and mutant banana Grande Naine plantlets. Plantlets were grown for 5 weeks on GA-containing or hormone-free media

Clone / Treatment	**Normal**	**Dwarf**	**Extra dwarf mutant**
Control (hormone free)	*) 14.2 b	10.8 c	9.6 d
GA_3 (10 mg/L)	21.7 a	14.9 b	9.6 d

* Values indicated by different letters represent statistical significance (P=0.05) using a multiple range variant analysis.

Another common variation is the phenotype of long and narrow leaf . This phenotype is also known as 'Giant' by growers.

Detection of Retro-transposons in Banana

We have attempted to detect expression of retro-transposable elements in leaves from Dwarf, Extra Dwarf and LNL plants that were induced to transcribe retro-elements by extensive duration of tissue culture. After twelve subcultures leaf samples were analyzed for the presence of mRNA encoding retro-elements. Using a specific set of primers we were able to detect a 300 bp retro-element in the LNL mutant.

The retro-element was only detected in the presence of the RTase in the synthesis of the cDNA and was not observed in any of the six wild type plants.

DNA gel blot analysis was used to estimate the complexity of retro-transposon enrichment of banana genomes following activation by tissue culture. Additional bands appeared in the lane loaded with the retro-activated mutant (LNL) DNA compared to that of the wild type. Furthermore, at least two bands that were present in the control show a significant higher label intensity as compared to the wild type.

DISCUSSION

Components of yield in bananas include number of bunches per mat, weight of each bunch and number of mats per hectare, which together determine the potential production in tons per hectare. In tropical areas, where plants grow continuously, the length of each growth cycle is an additional important component of yield, while in the Western Galilee, Israel, growers can only expect one growth cycle per year. In order to achieve high yield and better protection from adverse climatic conditions, producers tend to increase plant density by increasing the number of bunch bearing stems in each mat. Another important factor worth considering (mainly for exporting producers) is fruit quality, with emphasis on finger size and appearance of the fruit. Since yield has a positive correlation with all the parameters used in our initial selection, we predicted that selection for total yield in Israel would be useful in the tropics as well.

Although a narrow genetic base was sampled in our initial selection program, the quantitative and qualitative parameters were manifested in the subsequent field trials. The five clones selected performed above the mean values measured for the total population in all the relevant parameters: number of bunches per mat, average bunch weight and total yields per hectare.

The second step of clonal evaluation was performed on more than 120 bunch bearing plants per year, of each of the selected clones. The achievements of these clones in total yield, though differing from each other, were consistently above the control population.

The difference in performance between the third and fourth cycle can be attributed to the difference in climatic conditions.

During the winter season that preceeded the third cycle, the temperatures measured in the Western Galilee were unusually low. The residual effects of the harsh climate resulted in lower yields than expected for all clones including the control. However, even under the sub-optimal conditions of the third cycle, clones '17-'1 and '37-'5 produced approximately 17 percent more fruit per hectare compared to the control population. The traits that mainly contribute to yield are bunch weight and the number of bunches per hectare. Our data reveal a significant difference between the clones, in the context of these two parameters. While bunches of clones '6-6' and '5-1' weighed more than '42-5', the number of bunches per hectare for '42-5' out-performed the others in both the third and fourth cycle of growth. It is likely that a mixed population of several selected clones will be advantageous in a different ecological environment. Furthermore, to avoid deleterious effects due to a mono-culture population, we combined all five clones for evaluation in the Philippines. Propagation by meristem culture allowed an even mix of the clonal selections.

A crucial parameter for the banana export industry is fruit quality. Fruit loss due to inferior quality can reach 40 percent in some cases. To include this important characteristic in our evaluation, the yield for the Philippines was expressed as "box/stem ratio" in addition to "bunch weight" (rather than fruit weight per hectare). This ratio was significantly higher for the plants selected in Israel compared to locally selected clones. The difference reached an average value of 18 percent for the entire evaluation period and peaked during the most productive period around October 1995. Similar results were achieved with bunch weight.

The genetic clonal divergence supports the presumption of instability of the banana genome . Given that both clones (the one selected in the Philippines and that selected in Israel) clearly differed in various components of yield, shows that selection is influenced by environmental conditions. The geographic location of the Western Galilee allows selection unmasked by the effects of diseases and pests. Our data indicate that selection in the ecological conditions of the Western Galilee provides better results in ecosystems that involve biotic pressures. However, after several cycles of propagation, the Israeli clones seem to lose their advantage (unpublished data), probably due to a genetic drift caused by biotic pressure.

The limited gene pool of cultivated bananas for the export industry has led researchers to use alternative methods of germplasm improvement such as mutation breeding and recombinant DNA technology. However, since both of these techniques rely on small genetic changes, normally single genes, it is crucial to start with a good genetic baseline and with plants free of viruses. In this regard, our data suggest that pre-selected clones in marginal climates for banana production are advantageous.

Interestingly, the occurrence of "off-types" following extensive tissue culture is higher than expected from random cell mutation. However, most detectable 'off-types' exhibit either higher or lower sensitivity to GA. The results obtained by the GA assay on *in vitro* plantlets provide evidence to the hypothesis that 'Dwarf' and 'Giant' phenotypes are related to GA sensitivity. have shown that in addition to the difference in sensitivity, partitioning between the different GA metabolites differed between 'Dwarf', 'Giant' and the normal phenotypes.

Involvement of Retro-transposons in Somaclonal Variation

The high frequency of deviation from the original clone, occurring during the *in vitro* culture, suggests that the tissue culture process promotes genomic changes. However, the mechanism of somaclonal mutations in bananas is unknown. Recent studies provided clear evidence that extensive duration of rice cell culture activated retro-transposing elements. Furthermore, the tobacco retro-element Tto 1 undergoes transcriptional activation by wounding as well as by methyl jasmonate . The structural features as well as transcriptional activation of retro-elements resemble retroviruses. Under normal conditions they remain dormant and upon activation they are transcribed, reverse transcribed to cDNA molecules and reintegrated in new loci in the genome.

The DNA of the pre-cultured meristem and the mutants was analyzed by Southern hybridization using BR-1 as a probe. The probe hybridized to two fragments in the control (Fig. 7), while the LNL mutant hybridized to at least four additional fragments. The addition of bands on the Southern blot indicates a propagation of the retro-element in the LNL mutants. At this point, the number of integrated copies represented per band is unclear. However, in the LNL the hybridized signal was intensified. This may indicate multiplication of retro-elements in a close proximity to the original retro-sequence. A cascade form of retro-elements has been reported in other species . This phenomenon may explain the high frequency of a single phenotypic variation. If a gene associated with GA sensitivity resides in close proximity to the original retro-sequence and the distribution of insertions is biased to short distances, we expect a high rate of mutants involving GA sensitivity. We are currently investigating this hypothesis.

16

Plant Protoplasts

The development of a protoplast system permits regeneration of plants from a single cell. Hence protoplasts have become useful in most plant crops as a technique for their genetic improvement *sensu largo*, and can be considered as a technique complementary to conventional breeding Its application to somaclonal variation, genetic transformation and somatic hybridisation shows great potential. An exhaustive overview is given of the isolation and culture of protoplasts in banana.

INTRODUCTION

Protoplasts are naked cells that lack cell walls. They are spherical with a plasmolysed cell content and are contained within a plasmalemma. In principle, each individual protoplast can reform a cell wall, and later initiate either a callus through sustained divisions, or an embryo, defined as a somatic embryo. In banana they are obtained from *in vivo* tissues or *in vitro* cultures.

As early as 1902, Haberlandt stated that individual nucleated plant cells could convert into entire plants, either directly or through a callus stage. This phenomenon is termed totipotency and denotes the recovery of a whole organism from a single cell, and is also applicable to protoplasts. In theory all cells are totipotent, but in practice it depends on the past cellular environment. Usually the morphogenetic competence is retained at the unicellular stage that corresponds to a protoplast. However since protoplasts are not exposed to the stabilizing and inductive influence from neighbouring cells, they may have lost their plant regeneration capacity. In fact, factors such as the genotype or species of the manipulated plant, and the ontogenetic state of the explant source, exert a powerful effect on the regeneration potential of protoplasts. Consequently the development of appropriate *in vitro* conditions for protoplast regeneration is complicated. Notwithstanding this, successful efforts have been made in isolation, cultivation and regeneration of protoplasts, since Nickell and Torrey pointed out their merit for crop improvement.

Various tissues and an increasing number of plant species and genotypes have been successfully used in protoplast culture, but so far sufficient quantities of protoplasts for practical applications are not routinely met. While the value for agriculture of protoplasts still needs to be demonstrated, they are an invaluable tool for studies on permeability of ions and solutes [3-4], photosynthesis , phytohormones , phytochrome , and maintenance of totipotency . Moreover, protoplasts are useful for the uptake of foreign genetic material and to produce somatic hybrids through protoplast fusion. In addition, protoplasts are an excellent system for studies on cell genetics and even for plant virology.

These studies rely on isolated, clean and healthy protoplasts, which requires the appropriate choice of osmoticum, hydrolysing enzymatic solution, and pretreatment of the donor explant either *in vivo* or *in vitro.* This involves the right choice of explant age, cold treatment, phytohormone pretreatment, light intensity or photoperiod, and subculture rhythm, which affects both the internal metabolic status of the cells and cell-wall composition. Prior to cultivation isolated protoplasts need to be freed from enzymatic remains and debris, which are considered to be toxic.

Culture in or on solid medium is considered to be more advantageous than in liquid medium because the development of a single protoplast into a colony can be followed up much better. For development, different factors such as medium composition are crucial, especially the nature and quantity of growth hormones. Other important factors are the physical environment, plating density, and embedding conditions or the use of feeder layers, which all contribute in one way or another to cell-wall regeneration and cell division. Once cell divisions start, the level of auxin(s), as well as the colony density, usually need to be reduced to avoid overcrowding. This is done by subculturing, which gives the additional benefit that an adequate nutrient supply is maintained. Finally, for rooting, plants are usually transferred to cytokinin-free medium, possibly containing some auxin, and exposed to high light intensities, usually in illuminated plant growth chambers.

During recent decades, plant biology workers have recognised the potential of protoplasts in many experimental systems, since an efficient enzymatic method for protoplast isolation was first established by Cocking .

SIGNIFICANCE AND USE OF PROTOPLASTS IN BANANA

Banana is now easily amenable to *in vitro* culture, and plants are regenerated from various explants through organogenesis , embryogenesis, anther culture and even from cultured protoplasts. This has created opportunities for other biotechnological applications. Cell suspensions, for example, opened the way for gene transfer and improved the production, the quality, and therefore the manipulation of cell suspension-derived protoplasts. Through genetic engineering it has become possible to confer new traits on banana plants, using either particle bombardment or *Agrobacterium*-mediated transfer . Protoplasts facilitate the direct transformation of plant cells by DNA microinjection and electroporation .

However, many characters of agricultural interest are multigenic or ill-defined, and current transformation methodologies allow the integration of only a few foreign genes . Protoplast fusion, however, allows the transfer of several useful characters, even if detailed genetic or molecular knowledge of genes encoding for these desired characters is lacking. Protoplast fusion is therefore a complementary tool to increase nuclear and cytoplasmic variability and to confer desirable agronomic traits.

Important banana cultivars are susceptible to many pests and diseases, particularly *Mycosphaerella fijiensis* and *Fusarium oxysporum* f.sp. *cubense*, nematodes and insects. Very interesting traits of resistance have been identified in this genus for most serious diseases. However cross-breeding is very difficult in the genus since most edible cultivars are sterile even after profuse pollination by wild pollen-fertile genotypes. Protoplast fusion is therefore an option, as it can overcome such sexual barriers which occur frequently in banana, and which cannot be bypassed even through embryo rescue. Therefore somatic hybridisation between wild and cultivated banana is expected to produce hybrids combining agronomic traits with genetic resistance to pests and pathogens.

Protoplast fusion has the added benefit that it creates the possibility of generating asymmetric fusions , whereby selected cytoplasmic organelles, chromosomes or chromosome fragments from an irradiated protoplast donor could be combined with the genome of an acceptor protoplast. This strategy is of especial interest in banana, where the subspecies *balbisiana* is considered to be a source of multiple resistance and therefore could be partially utilised to improve banana cultivars, of which the most important component is the subspecies *acuminata.* Protoplast fusion also makes the creation of synthetic triploid banana genotypes possible by, for example, combining haploid protoplasts derived from anther culture with protoplasts from a diploid improved cultivar.

Chimerism and somaclonal variation are factors that seriously limit rapid clonal propagation of banana. By their nature, protoplasts avoid chimerism because they originate from a single cell, and protoplasts may be use to dissociate chimeric plants. On the other hand the cellular heterogeneity of protoplast populations can be useful for isolating somaclonal variants with improved characters such as increased yield or pathogenic resistance, so that somaclonal variation may also be exploited to provide new sources of genetic variability .

Banana protoplast isolation and culturing is nowadays routine, although protoplast research on banana started 15 years later than on model plants.

Now it is regrettably underused because of the strong emphasis on molecular and genomic studies in banana. Nevertheless, protoplasts have much to offer for non-conventional banana breeding, since they overcome sexual incompatibilities at interspecific and even at the intergeneric level, and allow the incorporation of multigenic traits such as yield, resistance to stress, pests and diseases. Finally protoplast techniques may improve banana when other classical methods have failed.

REVIEW OF WORK ON ISOLATED CELLS AND PROTOPLASTS

Early Experiments

The first attempts at banana protoplast culture were made in 1984 by Bakry. Various explants (leaf and sheath from *in vitro* banana plants, callus from floral explants, and bract and tepals from *in vivo* banana plants) of diploid and triploid banana plants belonging to the AA, BB, AAA and AAB groups were tested.

Enzyme solutions were composed of a modified CPW (Cell and protoplast washing solution) salt solution, containing 0.7 M mannitol, and a mixture of various enzymes at 0.1 to 5% (w/v) such as:

- Pectinases: Macerozyme (Yakult Pharma Industries, Japan) and pectolyase Y23 (Seishin Pharma Company, Japan);
- Hemicellulases: Hemicellulase (Sigma Chemical Corporation, USA) and Rhozyme (Pollock and Pool CTD, USA)
- Cellulases: cellulase R-10 'Onozuka' (Yakult Pharma Industries, Japan), Cellulysin (Calbiochem Behring, USA) and Driselase (Kyowa Hakko Kogyo Co., CTD, Japan).

Table: Cell and protoplast washing (CPW) solution

Component	Concentration (mg/l)
KH_2PO_4	27.2
KNO_3	100
$CaCl_2.2H_2O$	150
Mg $SO_4.7H_2O$	250
KI	0.16
Cu $SO_4.5H_2O$	0.025

The enzyme solution was adjusted to pH 5.6 with KOH and filter-sterilised (0.22 mm). Small explants of about 1-5 mm^3 were incubated for 9 h with shaking (80 r.p.m.) at 27°C in the dark. The mixture was sieved through an 83 mm metallic sieve, followed by dilution with KMC salt solution. After centrifugation at 100 *g* for 5 min, the pellet was diluted in a KMC solution. Whatever the plant material, enzymatic solution and plant preconditioning (light/dark), only isolated plasmolysed cells were obtained.

Positive results were obtained when nodular calli were used as starting material. Floral explants cultivated on MS medium complemented with 500 mg/l casein hydrolysate, 2 mg/l IAA, and 2 mg/l BAP lead to calli covered with nodular bodies. These calli were embryogenic. With the enzyme mixture 2.5% cellulase R10 'Onozuka', 0.2% hemicellulase, 0.3% pectolyase Y23 and 0.6% macerozyme, Bakry produced the first living banana protoplasts. After calcofluor staining, protoplasts were found to be cellulose-free, perfectly round in shape, variable in size and having a dense cytoplasmic content. Although the yield was quite low the protocol was reproducible.

This pioneering work was confirmed by Cronauer and Krikorian in 1986 (using proliferating shoot tips of the cultivar 'Lacatan' [AAA group]) and by Da Silva Conceicao in 1989. Later on regular progress was made by several scientists. Initially banana calli from protoplasts were developed in 1992 , followed by plant regeneration directly from protoplasts, protoplast transformation and somatic hybridisation.

Explants used as a Source of Protoplasts

In banana, *in vivo* explants are not a good source of protoplasts. Although Matsumoto et al. obtained protoplasts from *in vivo* bracts, in most cases initiation started from *in vitro* cultures. Leaf explants, slices of shoot tissue, roots or callus were the first choice as starting material for protoplast isolation because of their convenience. In fact, protoplasts can be obtained in banana from almost any tissue, including young leaves, sheaths, bracts, roots, and callus, but yields depend on the explant source.

Yields are quite low with most tissues and range from 0.1 to 2.8×10^6 protoplasts per gram, but the protoplasts are unable to divide. However, cell suspension-derived protoplasts gave high yields which ranged from 4 to 5×10^5 per gram with 'Long Tavoy'; from 1.1 to 3×10^5 per gram with *malaccensis*; 2 to 20×10^6 per gram with 'Maça' to 6.6×10^7 per millilitre of packed cell volume (PCV) with 'Bluggoe' . Essential was the embryogenicity of the suspension. The first sustained cell divisions in protoplast culture leading to calluses were observed by Megia et al. in 1992 with *M. acuminata* cell suspension-derived protoplasts. Megia et al. and Panis et al. improved the protocol. This was confirmed by Matsumoto et al. and Assani et al. who produced plants from such 'embryogenic protoplasts'. Consequently embryogenic cell suspensions are the starting material of choice.

Banana Genotypes Tested

Bananas are either triploid, diploid or tetraploid in order of importance for food production, but protoplasts have been obtained so far from triploid, diploid and haploid genotypes. Below are listed the cultivars and wild types tested; the letter A denotes the contribution of the subspecies *acuminata* and B the contribution of the subspecies *balbisiana*.

Triploid genotypes

- *Musa cavendishii* L.cv North banana (AAA);
- 'Lacatan' (AAA)
- 'Nanicao' (AAA)
- 'Bluggoe' (ABB)
- 'Maça' (AAB); 'Grande Naine' (AAA); 'Gros Michel' (AAA); 'Currare Enano' (AAB); 'Dominico' (AAB).

Diploid genotypes

- 'Pisang Lilin' (AA)
- *Musa acuminata* ssp. *burmanica* type 'Long Tavoy' (AA)
- *Musa acuminata* ssp. *malaccensis* (AA) [46]
- SF 265 (AA); IRFA 903 (AA); Col 49 (AA); 'Colatino Ouro' (AA); 'Pisang Klutuk Wulung' (BB); 'Pisang Klutuk' (BB); 'Pisang Batu' (BB) and 'Tani' (BB).

Haploids

These haploids were derived from anther-culture calli: either the callus © or leaves from regenerating plants obtained from haploid calli was the source (l).

- 'Pisang Klutuk Wulung' (BB) ©; 'Pisang Klutuk' (BB) (c and l); 'Pisang Batu' (BB).

Protoplast Extraction and Purification

Enzymes

Various enzyme mixtures are used depending on the author and banana explant (CS = Cell Suspension; P = Plant explant; C = callus). They are commonly composed of:

- Cellulases (Cellulase Onozuka R10, Cellulase Onozuka RS, Cellulysin, Macerase, Driselase)
- Pectinases (Pectinase; Macerozyme R10; Pectolyase Y 23)
- Hemicellulase (Rhozyme HP -150; Glusulase)

Protoplast isolation media

Different protoplast isolation media exist and are listed here:

- 0.5% Cellulysin, 0.5% Rhozyme HP-150, 0.125% pectolyase Y23, 100 mg $CaCl_2.2H_2O$, 10 mg $NaH_2PO_4.H_2O$, 58.5 mg MES 2-(*N*-morpholino) ethanesulphonic acid, 12.6 g sugar, pH 5.5;
- 1% Cellulysin, 1% Rhozyme, 1% pectinase, 0.3 M mannitol, 0.3 M sorbitol, pH 5.8 (P);
- 1% cellulase, 1% Macerozyme R10, 1% pectinase, 0.55 M mannitol, 3 μM MES pH 5.8 20-24 h ;
- 1% cellulase (Onozuka R10), 1% Macerozyme R10, 1% pectinase (Serva), 7 mM $CaCl_2.2H_2O$, 0.7 mM $NaH_2PO_4.2H_2O$, 3 mM MES, 10% mannitol, pH 5.8, 24 h;
- 5% cellulase RS (Yakult), 2% pectolyase Y23 (Sheishin Co.), 3% KCl, 0.5% $CaCl_2$, pH 5.6; 5 h ;
- 1.5% cellulase RS (Yakult), 0.2% pectolyase Y23 (Sheishin Co.), 0.2% hemicellulase Sigma Co, 3% KCl, 0.5% $CaCl_2$, pH 5.6, 5 h ;
- 1.5% cellulase (Onozuka RS, Yakult Japan), 0.2% pectolyase Y23 (Seishin Co.), 0.6 M mannitol, 0.1 mM $CaCl_2$, 0.5% PVP 40, 3.5 mM MES, 8 mg/l Bromocresol Purple, pH 5.7 15 h;
- 2% cellulase RS (Yakult Honsha Co. Japan), 0.5% Macerozyme (Sigma), 0.2% hemicellulase (Sigma), 0.25% pectolyase Y23 (Seinshin Pharmaceutical Co., Tokyo, Japan), 3% KCl, 0.5% $CaCl_2$, pH 5.6, 15-17 h ;
- 1.5% Cellulase RS (Yakult Honsha Co., Japan), 0.15% Pectolyase Y23 (Seishin Pharmaceutical Co.), 1% Macerozyme (Sigma), 204 mM KCl, 67 mM $CaCl_2$, pH 5.6, 12-14 h;
- The same as above but without macerozyme;
- 1.5% cellulase RS (Yakult Honsha Co., Japan), 0.15% pectolyase Y23 (Seishin Pharmaceutical Co.) 0.5%

Macerozyme (Sigma), 204 mM KCl, 67 mM $CaCl_2$, pH 5.6, 12-14 h.

It is standard practice to incubate the plant material in the enzyme solution for a digestion period of 1-24 h either with shaking (40 r.p.m.) or no shaking, generally at 27°C in the dark. The best source of protoplasts is from embryogenic cell suspension cultivated in the light (30 or 65 mE m^{-2} s^{-1}) or darkness and regularly subcultured. Initiation should start 3-5 days, 7-10 days or 2-15 days after the last subculture, followed by sieving through a 200-400 mm mesh just before enzyme incubation .

Protoplast purification

Protoplasts are purified from debris by sieving (100 mm and subsequently or directly through 56/32/25 mm sieves), which is preferred to flotation. With flotation purification, a 21% sucrose solution and centrifugation at 120 *g* is used.

Protoplast washing

After sieving, protoplasts are generally repeatedly washed to remove enzymes. Protoplast pellets obtained by centrifugation (50/66/90 g) are washed with protoplast isolation medium without enzymes ; cell suspension culture medium without growth regulator plus 10% mannitol ; or 3% $CaCl_2$ plus 0.5% KCl solution ; or 3% $CaCl_2$ 0.5% KCl solution and 10% mannitol; or 0.6 M D-mannitol, 0.1 mM $CaCl_2$, 0.5% PVP-40 and 3.5 mM MES.

Protoplast Quality Control

Complete cell wall degradation is confirmed both by the spherical shape of the released protoplasts and by the absence of fluorescence after staining with calcofluor white. Evans blue or FDA staining are commonly used for viability assessment.

Protoplast Culture Media

For banana protoplast culture, derived MS medium (supplemented with 2,4-D, or 2,4-D plus zeatin) or N-medium composed with N6 salt, vitamins, organic acid and sugar alcohol are commonly used. Due to difficulties in obtaining favourable protoplast development, many culture systems have been tested. Liquid and solid media (derived from MS) were initially tried for protoplast cultivation, but it was found that liquid cultures are not suitable for *Musa* protoplasts. In contrast, solid media supported sustained divisions.

To overcome the low growth response, other media were tested, such as semi-solid media, protoplasts embedded in solid or semi-solid media, nurse cultures, and nurse cultures with a feeder layer. Cocultivation in banana protoplast cultures was considered as it was reported that coculturing protoplasts of recalcitrant species, especially monocotyledons, with a reliable feeder culture, induced cell division. The banana protoplast literature describes various combinations of media and culture systems but, whatever the protocol, a high plating density (10^5-10^6/ml) was crucial , as well as the use of feeder layer , or nurse culture combined with a feeder layer. Protoplasts are cultured at 27 ± 1°C in the dark.

Culture in liquid media

For culturing protoplasts in liquid media, protoplasts with a density of 10^5-10^6 per millilitre are suspended in liquid media, with or without shaking. The following options are open:

- N_6 salt and KM vitamins, organic acids and sugar alcohols according to Kao and Michayluk, 6.3% (w/v) glucose and 4% (w/v) sucrose as osmoticum, 260 mg/l KH_2PO_4 and 100 mg/l 2-(*N*-morpholino) ethanesulphonic acid MES ;
- N_6 salts, KM vitamins, organic acids and sugar alcohols [50], vitamins of Morel, 0.4 M glucose, 117 mM sucrose, 0.5 mM MES, 1.9 mM KH_2PO_4 ;
- N_6 salts, KM vitamins, organic acids and sugar alcohols, vitamins of Morel, 0.4 M glucose, 117 mM sucrose, 1.9 mM KH_2PO_4, 2.28 μM zeatin, 0.90 μM 2,4-D and 5.4 μM NAA;
- MS modified medium, 5 mM 2,4-D and 10% mannitol ;

- 1/2 MS, 10% mannitol ;
- MS medium, 5 mM 2,4-D, 1 mM zeatin, 0.55 M mannitol ;
- MS minor nutrients, 1/2 major nutrients, MS vitamins, 10 mg/l ascorbic acid, 20 g/l sucrose, 5 μM 2,4-D; 0.55 M D-mannitol;
- MS preconditioned medium, 5 mM 2,4-D, 10% filter-sterilised medium from a one-week-old cell culture.

Culture on solid medium

For culture on solid medium (Figure 2B) protoplasts are directly plated at high density on a gelatinous medium or on a membrane which covers the gelatinous medium. Such media are composed of:

- MS minor nutrients, 1/2 major nutrients, MS vitamins, 10 mg/l ascorbic acid, 20 g/l sucrose, 5 μM 2,4-D, 0.275 M D-mannitol, 0.2% gelling gum; or
- MS 1/2 salt solution, 5% mannitol, 5 μM 2,4-D, 0.8% agarose or 0.2% Gelrite .

Culture in semi-solid medium

For culture on semi-solid medium protoplasts are cultivated on:

- 0.6% agarose, N_6 salt, vitamins of Morel, vitamins of Kao and Michayluk;
- 0.6% agarose, 0.35 M glucose, 40 g/l saccharose, 260 mg/l KH_2PO_4, 3 mM MES, and Pichloram, BAP, NAA, L-asparagine and L-proline at various concentrations ;
- Protoplasts at twice final desired concentration in culture medium, are mixed with the same medium containing 0.3/0.8% agarose, 5-10% mannitol, 0-10 μM 2,4-D .

Culture embedded in solid medium

For embedment in solid medium protoplasts at twice the desired concentration (2×10^6) are placed in a twofold concentrated culture medium: MS minor nutrients (× 2), MS major nutrients, MS vitamins

(× 2), 20 mg/l ascorbic acid, 40 g/l sucrose, 10 μM 2,4-D, but with 0.55 M D-mannitol. This is followed by mixing with an equal volume of solution containing a gelling gum at 0.6%, 0.55 M D-mannitol, previously melted and maintained at 55°C.

Culture in alginate or solid medium floating in liquid medium (bead type culture)

Protoplasts are cultured in a solid medium as in section 3.6.4, or in calcium alginate beads according to Schilde-Rentschler et al. For this purpose, the alginate solution (2.8% alginic acid, 0.4 M mannitol) is first autoclaved.

Protoplasts are added to the alginate solution to obtain a concentration of 1.0 × 10^6 protoplasts per millilitre. Later, drops of the protoplast-alginate mixture are polymerised in 0.4 M mannitol/ 50 mM $CaCl_2$ solution, and transferred into protoplast liquid culture medium (see below) in small Petri dishes. Pieces of solid medium containing protoplasts are transferred into liquid medium:

- MS medium, 10 mg/l ascorbic acid, 20 g/l sucrose, 5 μM 2,4-D, 0.275 M D-mannitol, with or without 0.3% activated charcoal;
- 'A medium' consisting of N_6 salts [49], KM vitamins, organic acids and sugar alcohols, vitamins of Morel, 117 mM sucrose, 1.9 mM KH_2PO_4 ;
- 'B medium' consisting of the 'A medium' containing 2.28 μM zeatin, 0.90 μM 2,4-D and 5.4 μM NAA .

Nurse Cultures

Living cells as nurse in liquid medium

Protoplasts can be maintained according to one of the following procedures:

- Nurse cell suspensions are maintained in N_6 salt with MS vitamins, 10 μM L-proline, 300 mg/l casein hydrolysate, 5 μM 2,4-D, and 3% (w/v) sucrose at pH 5.8;

- 1 ml PCV of nurse cells is freshly subcultured and mixed with 50 ml of MS minor nutrients, 1/2 major nutrients, MS vitamins, 10 mg/l ascorbic acid, 20 g/l sucrose, 5 μM 2,4-D and 0.275 M D-mannitol. Embedded protoplasts prepared as above (Sections 3.6.4 and 3.6.5.) are then immersed in the nurse solution. Cultures are placed on a rotary shaker at 120 r.p.m.

A Millicell™-HA (Millipore, 45 μm pore size) may be used as a physical barrier, to separate protoplasts from nurse cells during culture.

Feeder-layer Culture of Living Cells

For banana protoplast culture, *Oryza*, *Lolium multiflorum*, *Triticum monococcum* (Poaceae) and banana cell suspensions are useful as nurse cells. Feeder cultures are prepared one day before protoplast isolation by mixing 3 ml of feeder cell suspension with 100 ml of a medium containing 0.8% (w/v) melted Sea plaque agarose, MS basal salt, vitamins of Morel, 2 mg/l 2,4-D, 20 g/l sucrose, 72 g/l maltose, 250 mg/l glucose and 20 ml/l fresh coconut water . Feeder cells are then separated from overlaying protoplasts by a 10 mm nylon filter or 0.22 mm Millipore membrane.

Others options are:

- 2 ml of a nurse culture are poured on 20ml 1/2 MS, 5% mannitol, 5 μM 2,4-D, 0.2% Gelrite. The nurse culture is made from 1 ml of one-week-old suspension concentrated to 25% PCV mixed with 1 ml 1/2 MS supplemented with 10% mannitol, 200 mg/l myoinositol, 5 μM 2,4-D and 1.6% agarose .
- Nurse cells are mixed in melted medium MS (but with major nutrients at half strength), 10 mg/l ascorbic acid, 5 μM 2,4-D, sucrose 20 g/l, 0.55 M D-mannitol, 0.8% agarose (Sea plaque).
- 1 or 2 days before protoplast culture, prepare the feeder layer.

The cell suspensions are sieved through a 250 μm nylon filter to select small cell aggregates.

The PCM (Protoplast Culture Medium) liquid medium, which consists of MS salts, 9 μM 2,4-D, vitamins of Morel, 2.8 mM glucose, 278 mM maltose, 116 mM sucrose, 2.5 mM myoinositol (pH 5.7) is sterilised through 0.22 mm Millipore Millex GS filters (Millipore Corporation). Nurse cell suspensions are added to 100 ml double-concentration PCM culture medium to obtain a final cell concentration of 3-10% (v/v). 1.2 g agarose Sea plaque (Sigma) is dissolved in 100 ml water and then autoclaved (pH 5.7); when the agarose solution has cooled to 30-35°C, it is gently mixed with 100 ml PCM medium containing nurse cells, and 10-12 ml of the mixture is poured into small Petri dishes (5.5 cm diameter). After solidification, the medium is covered with a sterilised nitrocellulose filter (AA type, Millipore Corporation). 0.5 ml of the protoplast suspension in the A or B medium is transferred onto the nitrocellulose filter. Cell-wall regeneration is observed with calcofluor white. All cultures (liquid culture, alginate culture and feeder-layer culture) are maintained at 27°C in the dark.

Culture on preconditioned medium

This cultivation technique resembles the feeder-layer technique, except that the preconditioned medium is the cell culture filtered through a Whatman filter, followed by mixing the filter-sterilised medium with the solidifying medium .

Plant Regeneration

Sustained cell divisions are obtained with the feeder-layer technique, leading to plant development, while other techniques give only protoplast budding or few divisions. A high protoplast density (10^5-10^6 per millilitre) also appears to be crucial for plant regeneration. After 3-8 days of culture, first and second divisions are observed. After 4-6 weeks of cultivation, cell clumps consist of cells with embryogenic characters (i.e. small cell size, large nucleus with very stainable nucleolus, dense cytoplasm, small vacuoles). Since the clumps all have a similar size, it seems that they all arise at the same time from the protoplasts. These clumps exhibit a smooth surface, due to epidermis formation, and invaginate just like normal somatic embryos (Figure 3). These globules develop into rooted plantlets after 2-3 months.

For plant regeneration, cell clumps are transferred to a semi-solid medium without nurse cells, devoid of mannitol and growth regulators , or are transferred onto a medium containing 2.2 mM BAP and 2 mM pichloram for one week before transfer to MS medium containing 10 mM BAP to improve the rate of conversion of globules into plants . The membrane on which one-month-old clusters are growing is transferred onto solid medium with minor MS nutrients, 1/2 major MS nutrient strength, MS vitamins, 10 mg/l ascorbic acid, 20 g/l sucrose, 5 μM 2,4-D, 0.275 M D-mannitol, 0-2% gelling gum to produce embroids. Their maturation is achieved by transfer on hormone-free medium without mannitol, and finally on MS solid medium without growth regulators. Plantlet regeneration increased in the presence of BAP and IAA at low concentrations (2 mM) before transplanting to the soil in the greenhouse.

Protoplasts in a medium containing N_6 salts, KM vitamins, organic acids and sugar alcohols, vitamins of Morel, 117 mM sucrose, 1.9 mM KH_2PO_4, 2.28 μM zeatin, 0.90 μM 2,4-D and 5.4 μM NAA on banana feeder layer gives clusters in 3 weeks . Cultivation of these clusters on regeneration medium containing 2.2 mM BAP and 2.3 mM IAA results in somatic embryos with a 2% conversion rate of protoplast-derived cell suspension into embryos. These embryos are transferred onto MS medium supplemented with 1.2 mM NH_4NO_3 for rooting and elongation . Protoplast-derived bananas have now been field-established and are investigated for their performance.

PROTOPLAST HYBRIDISATION

Few groups have attempted somatic hybridisation in banana. The first experiments were made by Chen and Ku using PEG (polyethylene glycol) fusion, but at that time a protocol for plant regeneration from banana protoplasts was not yet available. Nowadays such research continues at EMBRAPA (Empresa Brasiliera de Pesquisa Agropecuaria) and at Paris-Sud University, France, using electrofusion approaches. This resulted in plant regeneration, and the first putative hybrids have been selected for ploidy by flow cytometry and established for field evaluation.

PROTOPLAST TRANSFORMATION

Banana transformation has much to offer for banana improvement, but technologies relying on protoplasts have not been fully explored yet. The only reports come from KULeuven (Catholic University of Leuven), Belgium and show that foreign genes can be inserted through electroporation. Embryogenic cell suspensions from the cultivar 'Bluggoe' were subjected to electroporation conditions consisting of 500 and 960 mF capacitance, PEG treatment, electroporation buffer, and heat shock, resulting in at most 1.8% protoplasts expressing the introduced DNA. These results show that non-chimeric transformed plants could potentially be produced from unicellular plant material.

PRACTICAL PROTOCOL FOR PROTOPLAST ISOLATION AND CULTURE IN BANANA

Isolation of Protoplasts from Cell Suspensions

For the isolation of protoplasts from embryogenic cell suspension cultures, suspensions should be selected 3-4 days after their last subculture. They then need to be sieved through sterile 500 and 200 mm stainless mesh. The enzyme solution is made of 1.5% cellulase RS (Yakult Honsha Co., Tokyo, Japan), 0.15% pectolyase Y23 (Seishin Pharmaceutical Co., Tokyo, Japan), 204 mM KCl, 67 mM $CaCl_2$ (pH 5.6) and centrifuged at 3000 r.p.m. for 1 h, after which the supernatant is filtered through a 0.22 mm sterile Millipore filter. Two millilitres of enzyme mixture is added to 1 ml of cell suspension, followed by overnight (12-14 h) incubation at 27°C without shaking.

Isolation of Protoplasts from Callus and Leaves

Embryogenic calli, or young leaves from *in vitro* plants (4-5 weeks old) are sliced and transferred into 150 ml Erlenmeyer flasks supplemented by a side nozzle connected to a Millipore filter (0.22 mm pore size, Millex GS filters, Millipore Corporation). Then 1 g of callus or leaves is mixed with 10 ml enzyme solution. For callus digestion, the enzyme solution consists of 1.5% cellulase RS (Yakult Honsha Co.), 0.15% pectolyase Y23 (Seishin Pharmaceutical Co.), 1% macerozyme (Sigma), 204 mM KCl and 67 mM $CaCl_2$ (pH 5.6). For leaf digestion, the enzyme solution consists of the same composition except for cellulase RS (1%), and the addition of 0.5 M mannitol. The mixture is incubated on a rotary shaker (30 r.p.m.) for 12-14 h at 27°C in the dark.

Purification of Protoplasts

For purification, the digestion mixture is filtered through a combined 100/25 mm metallic mesh to remove debris and large cell colonies. Protoplasts are then centrifuged at 66 *g* for 5 min. The pellet is washed twice (centrifugation at 66 *g* for 5 min) with a mixture of 204 mM KCl and 67 mM $CaCl_2$. Protoplast viability is then determined by fluorescein diacetate (FDA) and the yield counted on a Nageotte haemocytometer.

Culture of Protoplasts

To induce cell divisions in cultured protoplasts, nurse cultures have to be prepared a day before protoplast isolation. Cell suspensions in the exponential growth phase are sieved through a 250 mm metallic mesh to select only small cell aggregates.

The PCM medium is prepared by mixing MS salts, 9 mM 2,4-D, vitamins of Morel, 2.8 mM glucose, 278 mM maltose, 116 mM saccharose and 2.5 mM myoinositol (pH 5.7) and sterilised by filtration. Sieved cell suspensions are mixed with 100 ml double-concentration PCM liquid medium, to obtain a final cell concentration of 10%.

The subsequent steps are then:

- Dissolve 1.2 g agarose Sea plaque (Sigma) in 100 ml double-distilled water followed by autoclaving (pH 5.7). As soon the temperature of the agarose solution has decreased to 30-35°C, 100 ml of PCM medium containing nurse cells is gently mixed into it. Then 10-12 ml of the mixture is poured into small Petri dishes (5.5 cm diameter) and covered by a sterilised nitrocellulose filter (AA type, Millipore Corporation).
- Suspend 0.5 × 10^6 freshly isolated protoplasts in 0.5 ml liquid culture medium (1.0 × 10^6 protoplasts per millilitre) and transfer 300 ml aliquots onto a nitrocellulose filter.

The protoplast culture medium is composed of N_6 salts, vitamins, organic acids and sugar alcohols, Morel vitamins, 117 mM sucrose, 0.4 M glucose, 0.5 mM 2-(*N*-morpholino) ethanesulphonic acid (MES), 1.9 mM KH_2PO_4, 2.3 mM zeatin, 0.9 mM 2,4-D and 4.4 mM NAA (pH 5.7). Sterilise by filtration. The cultures are maintained at 27°C in the dark. Cell wall regeneration is observed with calcofluor white under a UV microscope.

Somatic Embryogenesis

Protoplast-derived embryos are individually picked up from the feeder layer after about 4-8 weeks of culture (depending on the genotype) and gently transferred onto regeneration medium containing MS salts, Morel vitamins, 88 mM saccharose, 2.3 mM IAA, 2.2 mM BAP, and 7.5 g/l agarose sea plaque (pH 5.7). They are maintained at 27°C in the dark. Regenerated plants are transferred onto solidified growth regulator-free MS medium with 1.2 mM NH_4NO_3 (pH 5.7) and cultured under a 16 h photoperiod (65 mE m^{-2} s^{-1}) at 27°C. After reaching a size of 10 cm, the plants are transplanted to the soil.

17

MOLECULAR UNDERSTANDING OF MYCOSPHAERELLA

A critical understanding of the molecular factors required by *Mycosphaerella* pathogens of banana to parasitize their hosts is needed to devise novel methods for control of these diseases. As a first step towards developing a molecular 'tool-box' to study pathogenicity and virulence factors of *M. fijiensis*, *M. musicola*, and *M. eumusae*, we developed a DNA-mediated transformation system, which confers on transformants stable resistance to the selective agent hygromycin B and constitutive expression of green fluorescent protein. These markers serve as a valuable means for detecting and monitoring transformed isolates *in planta* - on hosts or non-hosts - in the laboratory or greenhouse. We have also examined biological factors that may affect the ability of these fungi to infect their plant hosts. We provide results suggesting the presence of self-inhibitors of germination in vegetative spores, describe an assay used to detect effects of fungal phytotoxins on banana suspension cultures, and characterize fungal resistance to oxidative stress, specifically caused by singlet oxygen. As a result of these studies, we have gained a greater appreciation for the complexities of the interaction between *Mycosphaerella* pathogens and their banana hosts.

INTRODUCTION

Mycosphaerella fijiensis, the causal agent of the foliar fungal disease black Sigatoka, is the major worldwide constraint to banana and plantain (*Musa* spp.) production, and results in yield losses estimated at 33-76%. Yellow Sigatoka disease, caused by the closely related species *M. musicola*, has been largely supplanted by black Sigatoka in many banana-producing areas, but remains a significant problem at higher altitudes and cooler temperatures. The imperfect states of *M. fijiensis* and *M. musicola* are of the *Paracercospora* and *Pseudocercospora* types, respectively, although Stewart et al. have suggested reducing *Paracercospora* to synonymy with *Pseudocercospora*. Recently, a *Musa* pathogen with a *Mycosphaerella* teleomorph and an anamorph of the *Septoria* type was identified. This species, called *M. eumusae*, causes Septoria leaf spot, a disease with symptoms similar to those of black Sigatoka.

Previous studies indicate that the methods of pathogenesis used by *M. fijiensis* and *M. musicola* are broadly similar. The symptoms they cause and the structures of their conidia and conidiophores, as well as molecular techniques, are used to differentiate them. *In vitro* production of several fungal phytotoxins by *M. fijiensis* and *M. musicola* has been reported, but the role that such compounds play in the disease is unclear. Cultural, morphological, and pathogenic characteristics of the *Mycosphaerella* pathogens of banana are summarized in Carlier et al.

Knowledge of the mechanisms used by *Mycosphaerella* pathogens to parasitize *Musa* remains inadequate.

Our primary goal is to understand the mechanisms of pathogenicity and virulence of *Mycosphaerella* pathogens of banana at the molecular level. Towards this end, we have developed a genetic transformation system for these pathogens, and have used transformants expressing green fluorescent protein (GFP) to study and compare growth of the pathogens on a compatible host. We have also investigated other aspects of the biology of these fungi, such as self-inhibition of spore germination, the phytotoxicity of fungal extracts, and fungal resistance to oxidative stress. In this chapter, we document our progress in these efforts and discuss future directions.

RESULTS

Development of a Transformation System for *Mycosphaerella* Pathogens of Banana

Methods were developed to enable the transformation of *M. fijiensis, M. musicola* and *M. eumusae.* Isolates of each species were transformed with a construct carrying a gene encoding green fluorescent protein (GFP) and a gene conferring Hygromycin B resistance. Transformants of each species expressed GFP constitutively and were used to examine and compare infection processes on the susceptible banana cultivar Grande Nain. All observations were made with a fluorescence microscope using FITC fluorescence (excitation range 455-490 nm; 515 nm long pass barrier/ emission filter; dichroic mirror wavelength of 500 nm).

For all three species, the process of infection appeared similar, except that symptoms developed faster and were more severe on bananas infected with *M. fijiensis* and *M. eumusae* than with *M. musicola* strains. We hypothesize that this was because incubation conditions for *M. musicola* were not optimal, i.e. incubation temperatures were higher than typically found in natural infections. Leaf penetration was exclusively via stomata, often with the formation of hyphal swellings (stomatopodia) above the stomata prior to penetration. Hyphae grew between rather than into cells. All three species grew extensively within leaf tissue that eventually turned necrotic, at which point they grew saprophytically on the dead tissue. Additionally, leaf necrosis and chlorosis was often observed in advance of saprophytic growth of the mycelium on necrotic tissue, suggesting secretion of a phytotoxin. This work has been reported in greater detail by Balint-Kurti et al.

Growth of *M. fijiensis* and *M. Eumusae* on Non-host Species

Many fungi have been reported to form appressoria with equal frequençies on susceptible, resistant, and non-host plant species. However, in most instances, further fungal development is arrested prior to, or just after, entrance into the resistant or non-host plant. With the exception of some well-documented examples, plant non-host resistance remains poorly understood. It is clear from the published literature that several qualitatively different mechanisms are involved in non-host resistance, depending on the specific interaction.

To examine growth of *Mycosphaerella* banana pathogens on non-host species, plants of the tomato cultivar Moneymaker and the tobacco cultivar Sampson were inoculated with *M. fijiensis* and *M. eumusae* transformants expressing GFP. Conidia of both species germinated and formed stomatopodia over the stomata of both non-host plants. Subsequent stomatal penetration was not observed. However, abundant growth of an *M. eumusae* transformant was observed in a large necrotic spot on an inoculated tobacco leaf approximately 6 weeks post-inoculation. In the absence of observed stomatal penetration on tobacco, it is likely that this growth represented saprophytic growth on already dead or severely senescent tissue.

The ability of *M. fijiensis* and *M. eumusae* to locate stomata on the very distinct surfaces of banana, tobacco and tomato leaves indicates that these fungi do not rely upon host-specific topographical features of the leaf surface to guide them to the stoma, as is the case in some other plant-pathogen interactions. The failure of the pathogens to proceed beyond this point on the non-hosts is intriguing. Localized browning of tomato leaf cells directly under stomatopodia, which resembled a classical hypersensitive defence response, was observed at a few sites. However, on tobacco and in most cases with tomato, no such response was observed.

In the interactions documented here, it is possible that the pathogen needs a host signal, which is absent in non-host plants, to progress to the next stage of pathogenesis. Determination of the existence and identity of this signal would be of great value for understanding the pathogenicity mechanisms of these and other fungal plant pathogens. Alternatively, if one could demonstrate an active defence response in the non-host species, it might be possible to isolate genes involved in triggering this response from a genetically tractable non-host and use them to confer black Sigatoka resistance in transgenic banana.

Self-inhibition of Spore Germination

Spores of many fungi are inhibited from germinating when in a dense suspension or crowded upon a surface. Germination inhibition is readily lifted when the spores are dispersed. This phenomenon, known as self-inhibition of germination, ensures that spores do not germinate in the sorus or under conditions of intense intra-species competition. Germination inhibitors from several plant pathogenic fungi have been identified. Typically, these chemicals are small, variably water-soluble, organic compounds found on the spore surface. They are active at nanomolar levels and are often specific to a single species or to a group of closely related species. Genes involved in the production of germination self-inhibitors remain unidentified, and the mode of action of these self-inhibitors is not well understood. Liu and Kolattukudy recently determined that spore germination self-inhibitors of the rice blast fungus *Magnaporthe grisea* inhibit expression of the calmodulin gene, which is believed to be associated with appressorium formation by the fungus.

Initial *in planta* observations using transgenic conidiospores expressing GFP suggested that *M. fijiensis*, *M. eumusae* and *M. musicola* spores germinated well on the leaf when at low concentrations but that germination frequency was significantly inhibited at high spore concentrations. *In vitro* spore germination assays displayed a similar phenomenon. At high spore densities germination was inhibited, while at low spore densities this inhibition was lifted. We reasoned that, if a self-inhibitor existed, it might be possible to wash some or all of the self-inhibitor from the spore surface and produce higher frequencies of spore germination at high spore densities. Aqueous spore suspensions were prepared (~10^7 spores/ml) and divided into two aliquots. Half of the spore suspension was plated at serial fivefold dilutions while the other half was washed four times in large volumes of water and then similarly plated. The washed spores of both *M. eumusae* and *M. fijiensis* germinated at a higher frequency than the unwashed spores over a range of dilutions implying that a self-inhibitor had been removed from the spore surface. It is likely that some self-inhibitor remained on the washed spores, as they still did not germinate well at very high concentrations. These experiments suggest the existence of a self-inhibitor of spore germination in these species.

Light-activated Phytotoxins of *Mycosphaerella Musicola*

The asexual stages *M. fijiensis* and *M. musicola* are in the closely related *Paracercospora* and *Pseudocercospora* genera, respectively. Many fungal plant pathogens in the genus *Cercospora* produce the non-specific toxin cercosporin. When activated by light, cercosporin causes the production of activated oxygen species, particularly the highly toxic singlet oxygen. Although it has been suggested that cercosporin production is limited to fungi within the *Cercospora* genus, it is possible that related species produce toxins with similar properties. Cercosporin production has not been reported in *M. fijiensis* or *M. musicola.*

It has long been observed that the symptoms of black and yellow Sigatoka are ameliorated in plants grown in the shade, suggesting the production of light-activated toxins by these fungi. Indeed, there are several references in the literature suggesting that *M. fijiensis* and *M. musicola* produce toxins that are phytotoxic in *in vitro* assays though the exact role such toxins play *in planta* remains unclear. In collaboration with Dr. Jon Clardy's group (Dept. of Chemistry and Chemical Biology, Cornell University), we identified fractions of fungal extracts of *M. musicola* that were phytotoxic and exhibited light activation. Others have also detected light-activated toxins associated with these pathogens. It is only through the use of genetic techniques combined with analytical chemistry that we will definitively determine how any one toxin or family of toxins contributes to the virulence of the *Mycosphaerella* pathogens of banana.

Resistance of *Mycosphaerella* Pathogens of Banana to Oxidative Stress

Although the *Mycosphaerella* pathogens of banana have not been reported to produce cercosporin. We reasoned that if these fungi have the potential to produce light-activated toxins that could cause oxidative damage to plant cells, then they must also have the ability to resist such toxins. We tested isolates of *M. fijiensis* (seven isolates), *M. musicola* (two isolates), and *M. eumusae* (three isolates) for relative levels of resistance to eight photosensitizer dyes, following the general approaches of Daub. Each dye produces highly toxic, membrane-damaging singlet oxygen upon exposure to light. Not all fungi are resistant to singlet oxygen-generating dyes.

Conidia of each fungus were suspended in water at a concentration of approximately 250 spores per microlitre. For each isolate at each concentration of dye, an aliquot of 20 ml (approximately 5000 spores) was transferred to triplicate wells in 12-well microtiter dishes. Each well contained 1 ml Potato Dextrose Agar (PDA) containing one of the following photosensitizer dyes at various concentrations: Toluidine Blue (0,10,50,100 mM), Methylene Blue (0,10,50,100 mM), Erythrosin B (0,1,5,10 mM), Rose Bengal (0,1,5,10 mM), Phloxine B (0,1,5,10,100 mM), Eosin Y (0,1,5,10,100 mM), Cercosporin (0,1,5,10 mM), and Hematoporphyrin (0,1,5,10 mM). Appropriate volumes of organic solvent used for dissolving the dyes were added to control media.

Duplicate plates of each treatment were prepared and incubated in the dark overnight. We found it necessary to germinate the spores in the dark after plating, since direct exposure to high light levels at that time killed the spores. The next day, one set of duplicate plates was incubated at 20°C under cool white fluorescent light (70 $mE{\cdot}m^{-2}{\cdot}sec^{-1}$), and the second set was incubated in the dark at 20°C. All plates were examined for relative growth at approximately one-week intervals for 3 weeks. Plates were examined under a dissecting microscope, and at each evaluation, a qualitative rating of 0 (no germination), 1 (germinated spores), 2 (good hyphal growth), or 3 (extensive hyphal growth) was given. A rating of '2' or '3' suggested a degree of resistance at a particular concentration, although the growth rate may have been slowed in comparison with that on the same concentration of dye in the dark or on nonamended medium. Spores that germinated within the first week (rating of '1') but failed to grow thereafter were deemed sensitive to the dye at that concentration.

All isolates grew extensively at all concentrations of photosensitizer dyes assayed when incubated continuously in the dark. In the light, variability in relative resistance to the photoactivated dyes was observed between isolates of the same species. Interestingly, all isolates were resistant to photoactivated cercosporin at the highest concentration tested (10 mM), which is sufficient to inhibit the growth of several other unrelated fungi. In contrast, relative resistance to the other photosensitizer dyes was variable, depending on the isolate. Most *Mycosphaerella* isolates were as resistant as *Cercospora nicotianae* to the panel of photosensitizer dyes tested, with a few exceptions. Rose Bengal and Eosin Y were generally more toxic to the *Mycosphaerella* banana pathogens than to *C. nicotianae.*

Additionally, a few *M. fijiensis* isolates (e.g. Mf 294, Mf 435, Mf 301) were notably more sensitive overall to some of the dyes than other isolates of the same species. Most isolates were quite sensitive to Erythrosin B and Rose Bengal with the exception of Mf 436, Mf 303, and Mm 90, which showed resistance to 1 mM Erythrosin B, and Mm 90, which was resistant to 1 mM Rose Bengal. Although these data are not quantitative, we observed distinct variability among isolates of the same and different species in their responses to the panel of photoactivated dyes. Quantitative analyses, measured by enumerating the number of colonies surviving on each medium in both light and dark, might detect additional differences not shown here.

Oxidative burst, i.e., the release of active oxygen species (AOS), is one of the earliest responses in plant resistance to pathogen attack. Relatively little is known about the range of mechanisms that fungi use to counteract AOS-mediated plant defences. However, it has been well documented that bacterial pathogens of plants and animals can suppress AOS-mediated defences by secretion of AOS-scavenging enzymes such as superoxide dismutase and catalase. There is growing evidence that some phytopathogenic fungi have the ability to actively suppress AOS-mediated plant defences nonenzymatically by quenching AOS via induced synthesis of mannitol. It is likely that phytopathogenic fungi employ multiple mechanisms to defend themselves against AOS-mediated plant defences.

Our preliminary data reported here demonstrate that the *Mycosphaerella* pathogens of banana are generally highly resistant to several potent singlet oxygen-generating compounds, but that variability in resistance to such dyes exists. These differences among strains may suggest variability in the response of pathogens to plant oxidative burst. Can overall resistance or sensitivity to potent photosensitising dyes give us clues that might predict the virulence of a particular isolate? Specifically, does enhanced resistance to photoactivated dyes generally confer on the pathogen greater resistance to AOS generated by the host plant? Are pathogens that exhibit reduced resistance to photosensitizer dyes generally less virulent on their plant hosts due to a deficiency in the ability to quench plant-produced AOS? By extension, is there variability between banana cultivars in the oxidative burst generated in response to pathogen attack?

Three distinct genes that confer resistance to cercosporin have been cloned from cercosporin-producing *Cercospora* sp. *CRG1* (Cercosporin Resistance Gene 1) appears to be a membrane protein but has no strong homology to any gene in the databases. Under low stringency hybridization conditions, a *CRG1* homologue was not detected in genomic DNA of *M. fijiensis*, or of *M. musicola* and *M. eumusae* (P. Balint-Kurti and A.C.L. Churchill, unpublished data), suggesting that this gene is absent in these fungi. Therefore it is unlikely to play a role in resistance of the *Mycosphaerella* pathogens of banana to cercosporin or other photosensitizer dyes. *SOR1* (Singlet

Oxygen Resistance 1), now known as *PDX1*, is a novel gene involved in *de novo* biosynthesis of vitamin B6 (pyridoxine), which confers resistance to cercosporin and other photosensitizer dyes. Unlike *CRG1*, *PDX1* is present in numerous plants, fungi, and bacteria.

Table: Response of* M. fijiensis, M. musicola, *and* M. eumusae *to light-activated, singlet oxygen-producing dyes

CIRAD Isolate[a]	Photo-sensitiser[b]							
	TB(100)	MB(100)	EB(10)	RB(10)	PB(100)	EY(100)	C(10)	H(10)
Mf 294	R50[c]	S	S	S	S	R1	R	R5
Mf 301	R	R50	S	S	R5	R10	R	R
Mf 302	R	R	S	S	R10	R10	R	R
Mf 303	R	R	R1	S	R10	R10	R	R
Mf 435	R	R50	S	S	S	R5	R	R5
Mf 436	R	R	R1	S	R10	R10	R	R
Mf 722	R	R	S	S	R10	R10	R	R
Mm 90	R	R	R1	R1	R10	R10	R	R
Mm 318	R	R	S	S	R5	R10	R	R
Me 458	R	R	S	S	R5	R10	R	R
Me 485	R	R	S	S	R10	R10	R	R
Me 548	R	R	S	S	R1	R5	R	R
Cn[d]	R	R	NT	R	NT	R	R	R

[a] Mf = *M. fijiensis*, Mm = *M. musicola*, Me = *M. eumusae*. For each isolate, the same suspension of conidia was used for each dye tested

[b] TB = Toluidine Blue; MB = Methylene Blue; EB = Erythrosin B; RB = Rose Bengal; PB = Phloxine B; EY = Eosin Y; C = Cercosporin; H = Hematoporphyrin; the number in parentheses is the maximum concentration (mM) of dye tested

[c] R = resistant to the highest dye concentration (mM) tested in the light (indicated in parentheses at top of column); R# = resistant to the concentration of dye indicated and all lower concentrations tested in the light; S = sensitive to all concentrations of dye tested in the light.

[d] Cn = *Cercospora nicotianae*, data from [43, 53, 54]; NT = not tested

The *CFP* (Cercosporin Facilitator Protein) gene is related to members of the major facilitator superfamily of integral membrane proteins involved in efflux of and resistance to both bacterial and fungal toxins. Disruption of this gene caused a loss of cercosporin production in light-grown cultures and a 50% reduction in growth of *C. kikuchii* on medium containing exogenously added cercosporin. It is unknown whether homologues of *PDX1* or *CFP* are present in the *Mycosphaerella* banana pathogens. However, the presence of *PDX1* would not be surprising since it is involved in pyridoxine biosynthesis in many organisms. It has been suggested that the only cercosporin-resistant fungi so far identified are those that produce perylenequinone toxins such as cercosporin, phleichrome, altertoxin I, elsinochrome A, and hypocrellin A. Our results might then imply that the *Mycosphaerella* pathogens of banana produce light-activated toxins of this type. However, until such a toxin can be identified, we can only speculate on this topic.

DISCUSSION

The above examples from our own work suggest several areas of research related to pathogenesis or virulence of the *Mycosphaerella* pathogens of banana that would benefit from further analyses. The establishment of a transformation system for these fungi, in which GFP is used as a reporter gene, is an important first step towards developing the tools to dissect the mechanisms of pathogenesis used by these fungi. The next challenge is to identify and characterize specific genes involved in pathogenicity and virulence.

Several genetic approaches have been employed successfully to gain an understanding of pathogenicity and virulence factors in plant pathogenic fungi. Forward genetic approaches include mutant complementation and insertional mutagenesis. In mutant complementation, an untagged mutation is used to isolate a gene by DNA-mediated transformation and restoration of a particular phenotype. A complementation approach generally requires an efficient transformation system for the pathogen of interest. For fungi with low transformation rates, gene isolation by insertional mutagenesis may be possible. In this method, a mutagenic plasmid is introduced randomly into the fungal genome, transformants are screened for altered phenotypes of interest, and genomic DNA flanking the mutagenic plasmid is recovered.

Three approaches for insertional mutagenesis have been most successful to date and include Restriction Enzyme-Mediated

Integration (REMI) mutagenesis, *Agrobacterium*-mediated integration, and Signature Tagged Mutagenesis (STM). Each of these approaches has been reviewed by Gold et al. We have taken initial steps in adapting REMI for use in *Mycosphaerella* banana pathogen, but further work is required to make this approach useful. A critical component of any insertional mutagenesis approach is the availability of an efficient system for screening large numbers of fungal transformants for alterations in pathogenicity or virulence. This is not an easy task in the *Mycosphaerella*/banana system since symptoms are often not visible for several weeks after inoculation, and there are logistical problems associated with cultivating, maintaining, and inoculating large numbers of relatively large banana plants. The most efficient system will likely be based upon use of an *in vitro* leaf piece assay.

Alternatively, reverse genetic approaches could be used, in which specific genes predicted to be involved in pathogenesis are cloned by homology using hybridization with genes from other organisms or PCR using degenerate oligonucleotide primers. Multiple genes involved in critical signalling pathways of phytopathogenic fungi (e.g. MAP kinase, cAMP-dependent protein kinase, and mating pathways) have been cloned in this manner. Other types of pathogenesis genes that have been isolated using this approach are those for cell-wall degrading enzymes and gene families encoding chitin synthases. Alternatively, a protein of interest could be purified and the gene cloned by PCR, based on partial amino acid sequence or antibody screening. A necessary component of a reverse genetics approach is an efficient means for targeted gene disruption by homologous recombination to confirm gene function. Alternatively, homology-dependent post-transcriptional gene silencing techniques could be used. This approach has been of value in many organisms including animals, plants and fungi.

Finally, methods that rely on differential gene expression have been of great value in identifying pathogenicity and virulence genes in plant pathogenic fungi. In differential cDNA screening, subtractive cDNA hybridization, and more recently, suppression subtractive hybridization, it is possible to isolate genes that are specifically regulated during developmental changes that occur in the pathogen during infection, as well as at other times in the interaction of the pathogen with its host. Alternatively, differential display of mRNA, a PCR-based technique that generates bands visualized on sequencing gels, has been used to isolate genes that are expressed in phytopathogenic fungi only under specific growth conditions.

The *Mycosphaerella*/banana system is not an easy one with which to work, due to the slow growth of both the host and the pathogen and the difficulties associated with performing controlled infections. Black Sigatoka is, however, one of the most important worldwide plant diseases and this, we believe, makes the study of the interaction at a molecular level imperative. We hope that by pursuing some of the approached outlined here we can make significant progress towards this goal in the next few years.

Bibliography

Achuthan CR; Padikkala J; Jose Padikkala, (1997) *To stimulate the functional activity of the stomach, promote digestion and relieve pain.* Indian Journal of Clinical Biochemistry.

Burkill, H.M. (1985), *The useful plants of West Tropical Africa.* Royal Botanic Garden, Kew, UK 960 p

Chakraborty A., Devi R. K. B., Rita S., Sharatchandra Kh., Singh Th. I. (2004) *Preliminary studies on antiinflammatory and analgesic activities of Spilanthes acmella in experimental animal models.* Indian J Pharmacol.

Chaurasia O.P., Ballabh Basant and Raut B., *Indian Army and Green Governance: Conserving Medicinal Plants of Ladakh Himalaya,* Field Research Laboratory, Leh- Ladakh.

Hindu Folio, (2000), Indian Health Traditions, Source: FRLHT

Horrobin D.F. and Lapinskas P., (1998), *Plants for Food and Medicine,* Royal Botanic Gardens, Kew.

Jain S. K. (2000), *Global Resurgence of Ethnomedicobotany- The Indian Scene,* J. Trop. Med. Plants.

Joshi . S.P, 2000, *Medicinal plants,* Oxford and IBH Publishing Co.Pvt. Ltd., New Delhi.

Nandha Kumar R, Vishwanathan H, Suresh T, Mohan PS. *Antibacterial activity of Mappia foetida leaves and stem.* Bharathiar University, Coimbatore, Tamil Nadu.

Rajasekharan P.E. and Ganeshan S. (2002) *Conservation of Medicinal Plants Biodiversity— An Indian Perspective,* J. Trop. Med. Plants.

Taylor L, (2004), *Healing Power of Rainforest Herbs,* Square One Publishers Inc.

Index

□□□